Henna's Secret History

❀

Henna's Secret History

❀

The History, Mystery & Folklore of Henna

Marie Anakee Miczak

Writers Club Press
San Jose New York Lincoln Shanghai

Henna's Secret History
The History, Mystery & Folklore of Henna

All Rights Reserved © 2001 by Marie Anakee Miczak

No part of this book may be reproduced or transmitted
in any form or by any means, graphic, electronic, or mechanical,
including photocopying, recording, taping, or by any information
storage retrieval system, without the permission
in writing from the publisher.

Writers Club Press
an imprint of iUniverse.com, Inc.

For information address:
iUniverse.com, Inc.
5220 S 16th, Ste. 200
Lincoln, NE 68512
www.iuniverse.com

ISBN: 0-595-17891-X

Printed in the United States of America

This book is dedicated to my father Joseph Miczak. Without his old National Geographic Magazines and constant trips to museums, I doubt highly I ever would have penned a book such as this.

"Do come, O my dear one, let us go forth to the field; do let us lodge amoung the henna plants."

—King Solomon, Song of Songs 7:11

Contents

Acknowledgements xiii
List of Abbreviations xv
List of Contributors xvii
Chronology xix
Introduction xxi

Chapter One
   Henna by Any Other Name…is the Same 1

Chapter Two
   Pharaohs Gift? 23

Chapter Three
   Cleopatras Liquid Seduction 67

Chapter Four
   Beauty is in the Kholed Eye of the Beholder 86

Chapter Five
   The Lepers Only Hope 111

Chapter Six
   Aphrodite's Love Potion? 155

Chapter Seven
   From the Inside Out 167

Chapter Eight
   Black is Beautiful 180

Chapter Nine
A Small Liberation for Women *191*
Chapter Ten
Never Wear Henna to a Funeral…Unless Its Yours *220*
Chapter Eleven
Henna in Art *233*
Chapter Twelve
Camphire or Camphor? *243*
Chapter Thirteen
The Darker Side of Henna *261*
Chapter Fourteen
Vineyards of En Gedi *272*
Afterword 295
About the Author 297
Appendix 299
Notes 303
Glossary 307
References 309
Bibliography 319
Index 321

Front Cover: Painting of an Indian Queen with henna on her hands (part of Marie Anakee Miczak's henna art collection).

Illustration of henna by M. A. M.

Acknowledgements

❀

As always, I would like to thank my mother, Dr. Marie Miczak, who always takes the time to proofread my manuscripts and give sage advice. Her backgound in pharmacology and chemistry was extremly helpful in the writing of this book. In addition to the numerous contributers, I would like to thank those who passed their research on to me, including author Paul V. Beyerl whom provided a wonderful set of monographs on henna. I would also like to thank Renee Landkamer and Steve Schenk who passed on information. Both are fellow Suite101.com CE's. I would also like to thank Jennifer Wegner, Brian Pittman, Andrew Bentley, Hussein Elkhafaifi, Benjamin Foster, James A. Armstrong, S. Attar, Bruce Smith and the many others who were kind enough to answer my questions. Although I contacted many with requests for information, only but a few found time or had the knowledge to contribute; the ones who did were real gems. Thank you all!

List of Abbreviations

❀

BCE = Before Common Era (Before Christs Birth).

CE = Common Era (During Christs Life).

AD = (After Christs Death).

c. = circa.

g. = gram(s).

L. = *Lawsonia*

Lam. = *Lamarck*

List of Contributors

❁

Dr. Marie Miczak, D.Sc., P.hD
Dr. Miczak is a member of the American College of Clinical Pharmacology and the American Pharmaceutical Association. She attended Rutgers College of Pharmacy and has two doctorates, one in Nutriceuticals and the other being Nutrition Science. Her published titles include "Nature's Weeds, Native Medicine" (Lotus Press), "How Not to Kill Yourself with Drug Interactions" (Xlibris/Radom House), etc. She is frequently featured in magazines and newspapers including "Woman's World Magazine", "News from Indian Country" and "American Indian Review" to name a few. Her official website is http://www.miczak.com and she can be reached at 732-446-9669 &/or miczak@juno.com.

Nicolas Wyatt
Mr. Wyatt is the co-editor of the "Handbook of Ugaritic Studies" (Brill Academic Publishers) and works in the department of Edinburgh Ras Shamra Project (Ugaritic / Canaanite Studies) at the University of Edinburgh, Scotland. He frequently writes on the subject for magazines and journals.

Jonathan Warren
Mr. Warren is the Assistant Professor for the Department of English at York University, Canada. He was part of the Early Modern English Dictionary database project and author of the article "Reflections on an

Electric Scribe: Two Renaissance Dictionaries and Their Implicit Philosophies of Language".

Dr. Bala Subramaniam M.
Dr. Bala has done field and lab work on henna including extraction of oil from flowers, dye and testing of traditional Mehndi recipes for CAMPO Research (Pharmaceuticals), Singapore.

Andrew Bentley
Mr. Bentley is a professional herbalist who has studied traditional healing ways throughout Asia, Europe, and North America. In addition to research and clinical work, Bentley gives lectures on herbal medicine, acts as an informational resource for the media, the herb industry, and policymakers. email: Bentley@consultant.com web: http://www.rusticroots.com

Chronology

❀

Marie Anakee Miczak is also the author of:

"Secret Potions, Elixirs & Concoctions" (Lotus Press, 1999)

"Mehndi" (BBOTW/Infinity, 1999)

"How Flowers Heal" (Writers Club Press, 2000)

Introduction

❈

Like its contemporaries, coffee, tea and chocolate, henna is one of the most well known botanicals on earth. It is also quite possibly the most misunderstood. The henna plant, known by its Latin names of *Lawsonia inermis, L. alba, L. spinosa* and *L. ruba*, has up until this point been exclusively examined through its body adornment past. Commonly known today by the Indian Hindi name "Mehndi". What must be realized however is just like Swiss hot cocoa is but one form of chocolate, body adornment is but one use of henna. It is far from the "be all, end all" use. Few however have considered this and thus attempted to create a history which intrinsically revolves around its use for artistically staining the skin. This has thus left a "Swiss Cheese" of a historical accounting, which for many, leaves far more questions than answers. These questions apparently have no explanation in the world of body adornment.

This is what makes this book so unique and different. It looks at henna starting from the roots, quite literally. It only makes sense. Would you start documenting and researching chocolate's history from its creation as a bar in Hershey Pennsylvania? Of course not. Like chocolate, henna is a once living botanical which has been in use for thousands of years. To many it is medicine, a perfume, an aphrodisiac, wood for tools and was even used as a form of currency. Through henna's botany and medicinal usage's one will find its true, fascinating course through history. Whether it is truly medicinally valuable is not the point. If the ancients of the Old World believed it to be so, we have a basis for its real use. This is a starting point to see how many applications have feathered out through time. One can not plaster history with modern thinking. Doing so will denature the past. Surely mercury is toxic and has no medicinal benefits, yet it was used in orthodox medicine and will always remain a part of its history. Likewise we know today that henna has no beneficial effect on jaundice even though it has been used for such purposes for hundreds of years. Some natives of Africa still dip their newborns in henna in an attempt to prevent jaundice and not to simply ward off the Evil Eye as some note. Through looking at henna from a botanical standpoint, one can have a greater chance of catching discrepancies and recount in a realistic manner the many functions of henna, as a whole. It also gives a more complete picture which is filled with interesting narratives and insight . This elucidates not only this botanical but intimacies of life in the Old World.

A TORRID TALE

Henna is one of the most used botanicals of all time but the least written about and the hardest to research, exclaimed one professor I conversed with. "It is?" I thought to myself. Most likely because I veered off into the unknown by researching henna from a medicinally and botanically based standpoint that everyday brought a new, interesting piece to the henna puzzle that I was attempting to complete. Being skeptical has also been a part of my inherent nature…the child, teenager and woman who never ceases to stop asking "why?". Intensely inquisitive, I poured myself into researching this book and the writings of others before me. What I found is apparently few authors appreciated the power words and books have. A power to change history themselves, most notably in the form of incorrect recountings and inflated archeological evidence. This is not a new problem historically speaking. Take the Orientalist or Egyptologist from the Victorian era spewing a personal infatuation or hatred for henna into every archeological dig and /or discovery. The romantic fighting to hold onto customs in the midst of cultural independence and change. The so called, "New Age" enamored author attempts to add an esoteric meaning to every ancient daily practice. All have negatively colored henna's past. While many will point the finger at men misinterpreting the feminine use of henna as the root of all evil, women, especially in the 20th century, have done no better. Looking at suffragette Amelia Ann Edward's denatured henna poem or feminist Barbara G. Walker's strange henna lore, one can't entirely blame men. A few works from the 20th century written by male authors did seem to denature henna from that day forward. This thinking creeps into the majority of books on Mehndi being sold *today* and secretly perpetuates baseless claims and outright myths without detection. The major offenders being "Menhandi Rang Racni: A Folkloric Study of the Colorful Myrtle" by Dr. Mahendra Bhanawat which was published in India and "Mehndi" by a tattoo artist who neglected to use his legitimate name, published in the US. True detriment comes when one fails to examine claims, rushing ahead and stating them as fact and then for perhaps word economy, changes history all together with a summation. Case in point, the Baal and Anath epic, which is discussed further in chapter 2 of this book, goes from simply alluding to henna body adornment, which in itself is incorrect, to being documentation of an ancient "Night of the Henna" ritual for weddings in the Mediterranean. Of course if one asked for a copy of the archeological evidence of this ancient account of henna being used in regards to wedding ceremonies and was handed the Baal and Anath fragment from the 14th century BCE, one would instantly feel misled. Many however don't ask "why?" or for further evidence but instead follow right along, believing what they are told and then passing it on to others as bona fide fact. This is what has led to paintings in the Ajanta caves as being "evidence of henna use in India prior to the Mogul invasions" and the "cone", a British introduced item, to be called "the most ancient and authentic tool for applying Mehndi designs.". Myths like this chip away at the foundation of henna and ultimately the researching skills of authors whom continue to perpetuate them.

In the midst of the onslaught of "Mehndi" books, true historians, archeologists, Egyptologists and others in need of factual information pertaining to henna still echo "there is nothing on the market for *us* regarding henna's history…". If there is nothing for them, essentially there is nothing for the rest of us. Just because one is perhaps not a historian or a historically minded intellectual doesn't mean they deserves misconceptions and ambiguities intertwined with musings and myths. No work can be entirely correct. Even as the hand which writes it is imperfect, I believe if information is presented as it is *found* and allows one to make their own interpretations, it will teach, not mislead.

This brings us to the subject of plain untruths about henna, such as it will dye many colors naturally and fresh henna leaves create vibrant red Mehndi body art designs. Henna is very limited in what it can do as a dye, especially in the sense of cloth coloring. Contrary to popular belief fresh henna leaves *do not* dye vibrant red. In reality they dye yellow ocher; I know this because I imitated John Baptist Porta's work and did an experiment using fresh henna leaves. Such references could be simple mistakes or constructed untruths to cover a lack of evidence or prove a point, in itself wrong. Such myths are but a few of the *many* which are being circulated today. Thus, one must take into consideration why someone is compelled to document something such as henna in the first place. For me, I became extremely intrigued about henna as a *whole* after writing my 1st book "Secret Potions, Elixirs & Concoctions" and 2nd, "Mehndi: Rediscovering Henna Body Art". With a great love of *reading* I began to look for clear references of henna. What first started as personal interest blossomed into a work I hope young and old will enjoy and use to the fullest. Some however write for other reasons, such as for cultural preservation. This when done right is truly an important thing but when done for perhaps political reasons can turn into romanticized ramblings with little truth behind it. A fine example is a couple of books written in 50's India during its fight for independence *and* its bloody separation from Pakistan and Muslim Indian peoples. Specifically "Menhandi Rang Racni" by Dr. Mahendra Bhanawat and "The Art of Rajasthan: Henna and Floor Decorations" by Jogendra Saksena. While some of the information put forth by the male authors about henna is plausible in the context of India, a good deal of it is absurd, such as one quote, "henna is not a part of Muslim culture / traditions" or their implying non-Indians always had very crude designs. Henna went from being introduced by the Muslim Moguls to originating in India, coming directly from an Egyptian Pharaoh or even that India introduced henna to Egypt and the world. Not surprisingly the basses for these books was henna in the form of body adornment, specifically on Indian women. The essential oil of henna flowers and medicine made from its roots and leaves had already fallen out of favor, thus leaving a shell that body adornment is and always was henna's main function. This of course is very incorrect and when one tries to extract Mehndi body art history from the writings of Pliny Elder or King Solomon of the Bible, it simply doesn't materialize. It is because body art is not what is being discussed. Instead henna's rich perfume history may be recorded or its use as a currency. People long ago were not wasteful as we are today. Especially in the arid conditions of the Old World. In addition, prior to the advent of Western Medicine, health, foods, cosmetics, etc., were so intertwined that it is hard for us to comprehend today. Using henna singularly as a body adornment would not have been the mindset the ancients were in. Still, if one looks at henna as a whole and not through the cookie cutter shape of henna body art, one will be enlightened to the true history of *Lawsonia spp.* and its more cosmetic uses as well.

Flipping through the pages of this book, you may be highly surprised at some of the ancient uses of henna and information presented in general. However, when you look at the breakdown of henna, on a botanical standpoint one can see *why* it was used for what it was. This includes the Aspirin like properties which would explain its use for headaches and bug repellent actions which is why ancients would use dried leaves in bedding. Even more interesting is what the ancients themselves had to say about henna. From one of the oldest Pharaohs to a friend of Galileo, John Baptist Porta and into works of literature such as the stories of Sheherazade, one can get a grasp of both the importance and at times mundaness of henna. Being that the written word is used as a bases for this book and other evidence secondary, I believe it can help one to make their own conclusions. In addition I find it very important not act as Amelia Ann Edwards and others have through time by way of implanting words into the mouth of a writer. With only the exception of translating a local name for henna in brackets, I believe it is unfair to implant mentionings of henna where they are not really found and have taken every measure to prevent such from occurring in this book. Also taken into consideration is sometimes a number of plants are given the same title (i.e. Mignonette, Privet, etc.) and it can be hard to know for certain if henna is being mentioned or another botanical. In addition to the help from noted linguists, such dual meanings are noted in reference to historical recounting and supporting material is examined to gage if indeed henna is being recorded. Still, much importance was placed on obtaining correct localized names for henna as outlined in chapter 1 as many other "red" and "cosmetic" meaning names we know today such as ocher were *not* used in the Old World to denote *Lawsonia spp.*.

WHO WILL NEED THIS BOOK?

In addition to every woman or person hailing from an Old World local being a fine candidate, this book would be highly useful to a number of professional peoples. Most notably Egyptologists and archeologists wishing to find everything about henna in one place. One would not believe the number of peoples in such professions asking to be placed on a waiting list, in order to be contacted when this book would debut. In addition, medical doctors, botanists and herbalists would find this title very useful in the name of health. Whether it be preventing an adverse interaction or finding a new manner of treating an illness. Many titles covering henna on the medicinal level that this book does sell for literally hundreds of dollars. With the popularity of henna being applied to the skin, every dermatologist would be smart to have this book sitting on their shelf. Professors at colleges could use this as course text and this would be certainly a must read for anyone teaching about Mehndi body art and henna as a whole. Students working on theses and dissertations will find this book very helpful in addition to anyone interested in history, folklore and ecology. People wishing to grow henna in any capacity will also find this book extremely enlightening. Really there is something for everyone in this book, most likely because so many cultures and groups of people used henna in so many ways.

In addition to the interesting texts, many will also wish to use the references area to find the sources of information cited. One can be rest assured that the information presented from these sources is in accordance to what they themselves present, so one will not be dismayed at finding a surmised and thus denatured reference. The bibliography will point you to all sorts of further reading on subjects germane to henna's history but perhaps in themselves, need a full book to explain. This can include Dance Oriental and the application of Mehndi itself. For updates on the subject of this book and henna, one is encouraged to visit either http://www.mehndi.tajmahal.net &/or http://www.anakee.com. Since I have already penned a title on Mehndi or henna body art application, I will be leaving out such information found in "Mehndi: Rediscovering Henna Body Art" (Infinity, PA: 1999). If you are interested in purchasing the fully illustrated book, please see the reference section of *this* book, visit the aforementioned websites or call toll free, 1-877-BUYBOOK. Much to my delight, information in my book "Mehndi" is still pertinent and correct, even though I have found so much more material pertaining to henna as a whole for this book. I attribute that to doing my own research for the book and not basing it on the aforementioned books published in India or the US, as a number of other authors have seemingly done. This being so, I hope many will take the time to compare this book's *content*, with books presently on the market!

Chapter One

❁

Henna by Any Other Name...is the Same

The Many Names of Lawsonia inermy

In the language of flowers, Mignonette's hidden meaning was "Your Qualities Surpass Your Charms"
—Language of Flowers

Due to the fact henna has been in existence and used for thousands of years, there are a plethora of names associated with it. This, as you will soon learn later on, has led to much confusion about the uses of henna and its true history. Many times a single botanical was given two or three names depending on what stage of growth it was in. This has been found to be the case with henna. Henna that was to be powdered or used for dye would usually be harvested while still in a somewhat young, bush form. Henna that was grown to produce essential oil on the other hand, from the flowers, was allowed to grow into 20 foot trees. The small hedge/bush and tree form looked drastically different from one another. This may have caused those not familiar with the plant to incorrectly identify it in writings, especially recipes or attempt to create a new name for it. Also common was the use of one name for a whole species or grouping of "like" looking plants, Mignonette is a fine example of this. This was especially common before Latin names were used to separate each species, sub species, plant family and so on. The quest to document every plants medicinal value in Materia Medicas (Medicinal Herbals) early days (i.e. around 460 BCE) also helped with this problem. To make matters even more difficult however, henna the botanical is frequently interchanged with the art of body adornment / tattooing with henna &/or its essential oil form.

While it is unlikely that all the names can be listed in one place, I have collected the most common names given to *Lawsonia inermis* species and recorded them here. The ancient Arabic name for Henna or Hinna given to *Lawsonia inermy* is by far the most used and well known of all of the names. This name, originally "H*inna*" is believed to have been given to *Lawsonia inermy* by Arabic speaking Persians. The ancient Egyptian hieroglyphic name for henna is thought to be "Henu" and many will notice the similarity in the two. The root word for the Persian Arabic "hinna" has been suggested to be HNN meaning "tenderness". This made me wonder, henna isn't literally tender so why should its name be derived from such a root. I decided to ask the experts, as I most certainly am not a linguist. First I asked the Arabic linguist professor Samar Attar at Harverd University (Cambridge Massachusetts USA) about the root word of henna. At the beginning he considered the HNN—tenderness connection but stated he needed to consult his books and get back to me. The next day however he implored that I *dismiss* the HNN notion and stated the following:

"The *correct* root is: H(A)N(A)A') HNA with glottal stop at the end. Hannaa', tahni'a, tahni'ah: to dye someone with henna."

Now this to me made more sense, that the root word for henna would be in relation to dyeing. This was echoed by another linguistic professor of Arabic studies, this time from the University of Utah (Utah USA), Hussein M. Elkhafaifi, Ph.D. The professor also noted that hinna, meaning "to dye red" was indeed a word which originated in Arabic speaking Persia and is not thought to have been taking from another language such as Hebrew or Egyptian. Since Arabic is derived from Aramaic (or replaced it in Persia), this would tend to confirm the professors observation, as "kuphra" (KPR) and "alhinna" (HANAA) don't particularly resemble each other. Either way it would dismiss J. M. Allegro's theories that "al hinna" in Arabic was derived from Greek words that he not surprisingly linked to hallucinogenic mushrooms. Since "hina" is also henna's name in Farsi, it leaves me to wonder if the Persians really did formulate a name from the Egyptian "henu". Our English "henna" of today is taken from "hina" which remains *Lawsonia inermis*' most common and popular name. Since the henna plant itself doesn't appear to originate in Persia and very little is known about real ancient Persian use, it seems queries why a word they would use for henna would become so status quo. It would make more sense if peoples were already calling henna "henu" and the like in Egypt and then this name became transformed into the popular name we all use today. Certainly something to ponder! What is also very interesting though is the ancient Hebrew name / word for henna too has a root word meaning "to dye". Ko'pher is thought to be derived from the root word "cover" KPR which is related not only to dyeing but also the ritual anointing with fine oil or ointment. KPR however is today not thought to be directly indicative of henna itself and *numerous* professors and linguists were unable to locate for me notations to henna prior to the Biblical Hebrew Ko'pher in other Akkadian or Sumerian languages. This is also seen by J. M. Allegro's lack of Akkadian henna names as well. Apparently apt at finding ancient botanical names, especially in the Near East, he was only able to locate the Greek, Hebrew, Aramaic and Arabic names. Professor Benjamin R. Foster of Near Eastern Languages and Civilizations at Yale University (New Haven Connecticut USA) explained to me, after generously searching for Akkadian henna names:

> "From the best I can determine, henna is unknown in ancient Mesopotamia. This is surprising to me, but it may have appeared in the Near East much later. We think of coffee as being very Near Eastern, but, after all, it is a relatively recent import too."

This lack of an actual Akkadian (or Assyro-Babylonian) and Sumerian name or notations for henna gives more credence to biblical Jews being some of the first to spread henna about the Near East and also that henna originated in Egypt. It also means that pointing to henna use in *ancient* times, in such areas by Victorian authors (and others) must be taken with a grain of salt. Without an Akkadian name for henna, how are they finding it to translate? Akkadian was a very popular form of writing from 3000 BCE until the time of Christ. New discoveries also show an ever more, astronomically at least, advanced people in Mesopotamia so it makes little sense, if henna was so intrinsic to their society, that they would have no name and make no mention of it. Babylonian texts have recorded "sacred prostitutes" (priestesses) using pine seeds which Pliny states were mushrooms in rituals as well as other botanicals (some list over 200 documented), so the omission casts much doubt on henna use originating in this part of the Old World.

I noted that, to me, the Turkish word for henna, "Kina" sounded very much like the Arabic "hinna", so I decided to ask yet another expert. A linguist from England, Gareth Hughes, recounted that according to his research, the Turkish word "Kina" for henna was indeed derived or taken from the ancient Arabic or Farsi "hinna":

> "I've just leafed through the TDK Dictionary produced by Bilkent University: they definitively say that kIna comes from the Arabic root Hinnaa'! On reflection, the change in the first sound of the word could have been effected by mediation via Farsi."

In fact, in Farsi, the traditional language of Persia, henna is "hina", just like in Arabic. Obviously because it is taken from the Arabic word hinna. So what this appears to show is that henna was introduced to Turkey by the Persians or was traded heavily from Farsi speaking locals. What this also shows is a lack of connection between ancient Near East cultures extensively using henna in and about the area comprising Turkey. If they did, why adopt a word for henna from the Persians? Why not continue using their own word for henna? Some have attempted to suggest "chna" as being the root word of "kIna" thus making it somehow indicative of its use by ancient Cananite peoples. Linguist Gareth Hughes drew a blank, stating the word chna sounded awkward and unlike words used in ancient Semitic languages. Sure enough, the *only* reference to chna I found was in a very old book stating the word (spelling) is solely located in old Greek literature pertaining to the Phoenicians. According to the ancient Greeks, the Phoenicians used this word for the name of one of their male ancestors and for their country. A connection to henna was not made and since the ancient Greek word for henna looked and sounded nothing like chna, this appears to be something unrelated to *Lawsonia inermy*.

I find this all quite fascinating to say the least! Hopefully this will encourage other linguists out there to delve and find the hidden mysteries of henna's many names. Through language, I believe a lot can be gained and proper tracing can be done about henna's spread throughout the world. Below is a chart of popular ancient and modern names for *Lawsonia inermy*. More can be found in the Notes area, Section 3 of this book.

COMMON NAMES CHART FOR LAWSONIA SPP:

Ancient Egyptian: Henu / Henou / Ankh-imy [?] / Alcana
(NOTE: Some other ancient Egyptian names suggested include Puker / Tamrabene / Pouquer. The Puker and Pouquer at least appear to be from French translations of Egyptian hieroglyphics done in the early 20th century. I have personally reviewed the actual hieroglyphic letters used to spell henna and it translates as "henu" in the ancient Egyptian language. I have likewise not found mention of Pouquer, etc., in literature pertaining to Egypt or henna, so it is likely either a mistake or outdated historical accounting. If the author noting Pouquer, Marie Battesti, is anything like Ms. Edward, it would not be all too surprising.

Arabic: "H*inn*a" Henna / Hina /Al-khenna / Han (Given to the plant by the Persians) / Saumer / Hna (*Lawsonia alba* AKA: Black henna)

Sanskrit: Madayantika (Mendika—Mendhika—Medika) meaning "which gives color" (*L. inermis*) / Kurabaka (*L. alba*) / Tikta kasaya / Laghu ruksa / Zita / Katu / Kokadanta / Sahshara / Mendi-Mehndi (meaning myrtle)
(NOTE: Do to Sanskrits fragmented nature, there are numerous Sanskrit names used to denote henna. However these names can also be used to denote other plants/botanicals. Also, as will be seen in coming

chapters, there appears to be no real separation between using henna medicinally or on the skin to form body art designs.)

Latin / Latinate Botanical Names (Taxonomy): Alkanna / Lawsonia inermis (also L. inerma in older European texts) / L. alba / L. ruba / Lyzitus spina christi (natural henna leaf powder that has been bleached) / L. spinosa / L. mimata (perhaps a Japanese species) / Lawsoniaceae / Lawsoniear / Radix alcannae verae (henna roots) / Alcannae tinctoria (henna root dye)
(NOTE: Some believe henna's Latin botanical name, *Lawsonia*, is in recognition for John Lawson discovering and documenting it in India. Other (Taxonomy) records show however that the name *Lawsonia* was given to henna in the year 1753 which would have been after his death (c. 1711). John Lawson is most well known for coming to America in the early 1700's to survey the Carolina's and later disappearing. Hailing from Briton, it was said before coming to America, Lawson explored India and worked to give plants Latinate names. Unfortunately not much is documented about him before his exploration in the Carolina's and book "A New Voyage to Carolina". One thing to be noted is that according to books about 18th century British chemical terms, henna is listed under its Latin name of *Lawsonia inermis*, which gives further indication that the linkage between the English John Lawson and henna may be founded. Especially since it wasn't a Dr. Lawson who discovered Lawsone (the main component of henna) as some claim. Incidentally, "*Inermus*" means "without thorns", indicating henna has smooth bark and "*alba*" means white, in relation to its white flowers. The epithets "*spinosa*" means "with thorns" and "*ruba*" denotes the color red in relation to henna's sometimes red blooms. In the mid 1800's *Lawsonia* was placed under the Family of *Lawsoniaceae J. G. Agardh* for which it is the only member.)

Ancient Greek: Kypros / Kupros (meaning Henna Tree) / Cypri / E-ti [?] / Cyprus / Appelle cyprus

Ancient Hebrew: "*ko'pher*" Copher / Koffer (Biblical) / Eshkol hakofer / Bapar [?] / (NOTE: Today commonly known as Henna.)

Ancient Aramaic: Kuphra (NOTE: Obviously derived from the Hebrew ko'pher.)

Latin: Cypri

Chinese: Wu-Bai-Zu / Wu-Pei-Tzu (meaning Gall of Henna) / Zhi-jia-hua (meaning Fingernail Flower) / Ji-xing-zi (henna seeds / berries) / Hai Na (meaning henna plant and most likely derived from Persian Arabic Hina / Henna)

Dialects / Areas of India:
Hindi: Mehndi / Hena
Bengali: Mehndi / Shudi
Rajasthani: Mendi
Marathi: Mehndi / Mendhi
Gujarati: Medi / Mendi
Tamil: Marithond / Maruthani / Kuraci / Mayilainanti / Kurantakam / Kurici / Kurantam / Korantam (Latter 5 *L. alba*) / Marutonri-hinna / Marutani / Marutonri
(NOTE: Tamil is a Dravidian language meaning it is unrealted to any other known lanuage including Sanskrit, for more names read the Notes area Section 3)
Telugu: Goranti / Maidhakku / Gorata / Goranta (latter 2 *L. alba*)
Kannad: Mailanci / Gorante / Mayilandi' gorante / Gorate / Kurunta / Kurige (latter 3 *L. alba*)
Malayalam (Malyalam): Pontalis / Mailanci / Mailanci' pontalis
Oriya: Benjati / Manjuati

Kashmiri: Mohuz
Punjabi: Mehndi / Nakrize / Panwar / Mehandi
Mundari: Mindi / Bind / Mindi' bind
Kanarese: Gorante / Goranta / Korate
Kerala: Mayilanchi
Telegy: Goranti
Pakistan: Korinta (*L. alba*)

Island Names:
Malaysia: Inai / Berinai
Cyprus: Kypros (noted by Homer)
Sir Lanka: Marathondi / Maratoli [a Dravidian lanuage]
Philippines: Cinamomom / Cinamomo Del Pais / Kolinta / Korinta / Kulanta / Kuranta (latter 4 is Tagalog)
West Indies & Jamaica: Reseda (originating from the name Resado meanig "to calm") / Mignonette (pronounced "y*u*net" and meaning "little darling" in French)
Java: Pachar kuku
Japan: Tsume hana

Other Common Names:
Mendy
St. Francis' Mignonette
Alquena
Jamacia Mignonette
Fouden (Africa)
Mignonette Tree (Victorian)
Faria (henna flower / fragrance—Arabia)
Al-henna
Meti
Beberiska (Morocco—the blessing effect of henna)
El-henna

Mandee
Orcanete / Alcanete (French)
Egyptian Privet
Kurabaka (Sanskrit)
Medudi
Quene (French)
Bhurara
Alcaneta (Old Spanish)
Smooth Lawsonia
Gol henna' (henna flowers—Iran)
Han (Africa)
Chinne
Hina (Farsi)
Mendi (Pakistan Sind)
Panna
Muhanoni / Muheni / Hanuni (African Swahili)
Hinai
Lhenna / Lhenni (Berber)
Hen'na adj
Hna
Kena / Hennastrauch (Croation)
Heena Tree
Alkanna (Italian)
Alcana
Alhena (Spanish)
Henne
Hanabandan
Gulhina
Krapeu (Cambodia)
Alcanna
Kina (Turkish)

Alchena
Lalle (Nigeria Hausa)
Reseda de cayenne (French Guiana)
Simru
Tamra-henni (Arabia)
Oleum Cyprineum (Oil of Henna, Cleopatras henna perfume)
Marudani
Alquena (Spanish—Spain)
Jan Chih Chia Ts'Ao
Iswan (Belgium)
Henne suractive
Egyptian Ligustrum (Egyptian Privet) / Alcharma (Henna in Segond's translation of the Bible)
Tche Kia Hoa (China)
Henneh (Egypt)
Iplik kinasi (Turkey)
Henne noir
Alchanna (Medieval Latin)
Khenna
Lele (Africa)
Henne naturel inde
Shazab (Yemen)
Menhada (India)
Hana
Inni (Baluchistan)
Danbin / Dan (Burma)
Inai parasi (Sumatra)
Sahshara [?]
Alhea
Yoranna [?]
Tien Kao (Thailand)

Flower of Paradise
LAIN5 (USDA Symbol for *Lawsonia inermis*)
LAAL7 (USDA symbol for *Lawsonia alba*)
C.I75480 (Natural Orange #6)
LAWS (Family abbreviation)

> **Henna** (Hen"na) *n.* Arabic *hinna* alcanna Cf. Alcanna, Alkanet, Orchanet. (As appeared in early 20th cebtury Websters [TM])

The mistake some make (and many have made in the past), is to look for ancient names meaning "red" or those that have to do with makeup/cosmetics in general. One thing to understand is henna was not the only red nail color or hair stain in existence. Other substances were also used for nail color, especially in colder and mountainous areas were henna does not grow well. Vermilion (HgS), a crystalline form of mercuric sulfide (also known as cinnabar) is one example and would *not* be indicative of henna. Various berries were also used and so was red ochre (Fe2O3xH2O), a type of natural clay. Being that this is so, it is important to find names and notations directly related to the actual henna plant and not the results it may produce. Also if cultural groups were using vermilion, red ochre, etc., it does not automatically denote henna use as well.

It is also important that some translate the Sanskrit *L. inermis* name to Raktgarbha, Ranjaka, Ragagarbha and Ragangi. These names have been said to be referring to Blue Indigo (Nil-awari) instead in older texts but since Indigo's true Sanskrit names all start with "a" (ajara, avari, ajura, avuri, etc.) this may not indeed be the case. The shrubs similar appearance can cause a mix-up . The above names sound very similar to ones used for turmeric and in early texts could really be denoting the golden spice. In some areas however, in India, henna is known by these names orally. Henna actually has many names and interpretations in Sanskrit due to the fragmentation of the ancient writing system including "Kokadanta". Especially in India, the henna plant is separated into the two main colors of flowers produced. Madayantika for the white to yellowish flowers and Kuranaka for the red, pink and deep purple colored flowers. Ancient Indian scholars also mistakenly attributed henna as a type of Barleria and Rue hence why some of the Sanskrit names are translated as "henna rue" and "henna barleria" etc. Even though henna is not a direct species of the aforementioned, the Sanskrit names can be indicative of the henna species *Lawsonia alba* which is thought of as inferior in most cases to red henna. These include the names "Aravanam", "Marutonri" and "Aivanam". Some suggested Sanskrit names which proved incorrect included "Timira" which actually means "dimness of the eyes" and "Yavaneshta" which is a synonym of "Neem", both showed no connection to henna.

Another correct name I've found for henna is "Cyprus" which some confuse with the Cypress tree. John Gill wrote "The Cypress tree is reckoned by Josephus among the odoriferous trees which grew about Jericho near to which Engedi was." He also explained "alhena" (henna) was the cypress tree which is of course incorrect. Cyprus is the newer spelling of Homer's Kypros meaning henna plant/tree in relation to the Island of Cyprus. Homer (750 BCE) noted that henna grew extremely abundantly on Cyprus, hence its given name. Prior to Homer's remarks, the Island was known by a number of other names, which is discussed further in chapters to come. Some believe that "Kypros" is an ancient Greek translation of the ancient Hebrew word / name for henna, "Koffer" (Ko'pher). The similarities certainly *are* there. Others however believe kypros comes from the ancient Egyptian henna containing incense blend of "Kyphi". Kyphi is a Greek translation of the Egyptian name. The Greeks were extremely fascinated by Kyphi (much to the horror of the Egyptians), as an aphrodisiac, and the ingredients used.

Mignonette (not to be confused with the French name of a sauce) on the other hand is a more European, especially French, name given to henna and is constantly found in poetry. In 1853 Mary Chauncey writes in her book "Mignonette—a very fragrant Annual from Egypt" and writes that its hidden Victorian meaning is "moral and mental beauty". Later that meaning was changed to "Your qualities surpass your charms" by various flower meaning books including "Language of Flowers" by Kate Greenaway (1885). The meaning was then changed again to "worth and loveliness" at the turn of the century. The only problem is Minonette is used to denote a number of plants. Henna however is the only one sturdy enough to be turned into the classic and highly aromatic Mignontte Tree.

Returning to the subject of incorrect name associations, the ancient Egyptian "kiki" is not referring to henna. It refers to 1.) a very large gourd that grows in Egypt and 2.0 oil of castor or the castor-berry (*Ricinus communis*) from which the oil is extracted. The oil which Pliny Elder called "cici" was used in ancient Egypt as a fragrant lamp oil. Another name suggested for henna is the ancient Greek word "Kino" (perhaps because it sounds like Kina) which actually means "red juice of plants". It also means "The dark red dried juice of certain plants, used variously in tanning, in dyeing and as an astringent medicine.". Since fresh henna does not produce a red juice but instead a yellow liquid which turns brown when dried, it's unlikely this was a definitive name the ancients used to denote *Lawsonia inermis*. Another reason why it would not signify henna is because according to a number of botany sources including "A Modern Herbal" by the gifted herbalist Mrs. M. Grieve, "The term Kino is applied to the juice of plants inspissated without artificial heat". Otherwise, the juice was dark red as it exited the fresh botanical...unlike henna which needs hot liquid and acids to bring out its reddish brown color. Just as noted above, its fresh juice is yellow. Although there are a number of botanicals that are called Kino, *Pterocarpus marsupium* is thought of as the classical or real species. High in tannic (kino-tannic) acid, its known as the "Malabar Kino" and "Basterd Teak". The juice known as Kino as well, is obtained from cuttings made in the trunk which allows it to ooze out. The juice dries brittle, jewel like pieces which are black-red in color and when placed in the mouth or water, turn blood red and produce kino-red dye. Also one must ask themselves why Theopharastus, Pliny Elder, Galen and others didn't use this name to refer to henna. Its quite clear, at least from the words of Dioscorides that the ancient Greeks and Romans knew of henna's staining and coloring abilities. The high tannic acid content would explain Kino's use as a tanner and dye in addition to its medicinal use for various infectious conditions. Even though their are a number of tree's which Kino can be obtained from, I have never seen any indication of a linkage to henna.

While Alkanet isn't necessarily an incorrect name, as it is derived from Alcana and Alchanna, which both mean "henna", it is the root taken from a number of plants in the *Anchusa spp.*. There is a great number of plants which fall under the name of Alkanet, including at times real henna roots. There could be a number of reasons for this, which are explored further in other chapter of this book including chapter 12. One of the reasons for the confusion is henna's roots have been long used by pharmacists to color oils and medicine and was given the pharmaceutical name of "radix alcannae verae". Henna roots seemingly were used as a substitute for Alkanet or buglos roots in many parts of the world, although less so in Europe because it was expensive. The German author C. H. Ebermayer wrote in the 1821 "Handbook of the Pharmacist" the following:

"The root of the *Lawsonia* of the East (*Lawsonia inermis* L.), which is also known under the name of "redix alcannae verae" also contains a red, soluble principle coloring in oils and alcohol."

The ancients however seemed to separate the two. For example, the ancient Egyptian name for Alkanet was "nesti", as opposed to "henu" for the henna plant itself. It also may be that the root which dyes bright red, was likened to henna, thus the resulting name, mentioned by Hipprocrates (400 BCE), "The black chameleon, when prepared with the juice of the fig. It is to be prepared roasted and Alkanet mixed with it" and by Pliny (60 AD) whom explained it was used to color rose perfume. Henna and Alkanet seems to have a strong connection. Henna and Alkanet are in fact distantly related to each other as henna's family of plants fall under Lythraceae and Alkanet under Boraginaceae which both fall under the division of Magnoliophyta and class of Magnoliopsida. I have seen Alkanet and it doesn't resemble henna as the flowers slightly differ in color, efflorescing in shades of bright blue which changes to purple. The ancients always describe henna flowers as being white, yellow, pink or red in color, never blue. It may be noted that Alkanet too can have yellow and white flowers. What is foggy is the name Alkanet itself, which is derived from the Middle Age English and Spanish words directly related to the Arabic "al-hinna" meaning "the henna" and "hinna" meaning henna as well. The Latin name for Alkanet, *Anchusa* however is derived from the Greek word "*anchousa*" which simply means "paint". Paint of no specific color. Alkanet, unlike henna is also found in Europe (although not native) and has been used for millenniums to color culinary dishes and as a cosmetic. One thing very interesting though is that Alkanet has been used, like henna, for ages to stain the nails and skin in a real red color. This practice was especially carried on in India where it is used as a substitute or additive to henna at times of festivities.

Speaking of India, many wonder where the word "Mehndi" came from and what it means. A simple question that until now no one has bothered to answer. The reason for this is explored in this books chapter on Mehndi body adornment using henna, chapter 9. Many will note that even though it is today thought of as the art-form in which henna is used to stain the skin, various areas of India refer to the henna plant by Mehndi and that name only. This is because "Mindi" and "Mehndi", coming from the ancient Sanskrit, especially in the Rajasthan area, means myrtle. It is also a shorter version of the Sanskrit "(Mendi)ka". Originally henna was known as "Hina Mehndi" but then the "Hina" appeared to be dropped and "Mehndi" used singularly. It would also explain why Mehndi is thought to have originated in the dialect of Hindi, as Hindi is thought to be a direct descendent of Sanskrit. In parts of India the henna plant is thought of as a simple "Myrtle", which encompasses a number of plants. This is in keeping with the henna being grouped with other species under Alkanet, Privet, Mignonette, etc. It however does not mean that references to myrtle in other parts of the world automatically denotes henna. This is seen by Israel and the Torah's view of myrtle as a "thick tree" (Etz Avot) whose *leaves* were said to have been used by the ancients to line coffins and fumigate areas. Most myrtle's produce aromatic oil in their shinny leaves which stayed apparent for days, thus the use in houses and graves to mask odors. Myrtle became a symbol of longevity and prosperity due to its long lasting scent; branches of leaves being frequently used at weddings and other gatherings throughout the Middle East. The biblical Hebrew name for myrtle was "hadhas" for which Queen Esther was named after; her Hebrew name being Hadassah meaning myrtle. It's quite obvious the biblical Jews knew about henna (koffer) and since they and other cultures made no association with it and myrtle, one must be careful not to intertwine the history of these two very different species. Some however would like to blend the histories, as myrtle has much esoteric and mythological associations, where henna seemingly does not.

I will touch lightly on the subject of the ancients confusing myrrh and henna, only because it has been suggested. In addition to there being no evidence of this, myrrh, like frankincense, refers to the end product which is dried, golden-brown resin pieces. The ancient Egyptians which called it "khery" and "antiu", among other names, do a nice job of summing up what it is, "its color is like gold, it has many grains its egg-shaped grains are like the egg of the swallow.". The aforementioned comes from inscriptions found on ancient Ptolemaic temple walls. The Egyptians appeared to have separate names for the plant which myrrh was harvested from, which was said to be "nuhi-heri". The leaves and other portions of the tree (*Commiphora spp.*) do not seem to have been used and also does not resemble henna. Incisions to the trunk or stems is not needed in order for resin to ooze out but in situations of harvesting, such are made to increase the flow. When drops of resin fall to the ground, they become hard and extremely aromatic. In Roman times, myrrh was added to wine for an extra intoxicating effect, which was called "Murrhina". Perhaphs the "hina" portion of the Murrhina was the confusing factor to cause some to link myrrh and henna together. Gareth Hughes was kind enough to give some further information on the subject:

> "Hebrew has a number of words for myrrh, but they are quite difficult to differentiate. The two main ones are lot, and mor/môr. I think that môr is the main word for myrrh as the Ugaritic root is 'mr', and the Akkadian is 'murru'. Lot seems to refer specifically to the resin from the cistus or 'rock rose' (*Commiphora villosus* or *Commiphora salviaefolius*), which is common in Palestine. The Hebrew word nešeq may be something to do with myrrh, but I think that is very unlikely.

The English word myrrh comes from the Greek mýron, but mýron means 'ointment' in general. The Greek word for myrrh is smýrna (I must confess this looks similar, but I don't have the knowledge of Greek etymology to say so). Smýrna is also the name of a prominent Ionian city on the Ægean (modern Izmir)."

Since henna does not produce abundant resin, there is no reason to confuse the two botanicals. Also the names given to myrrh don't appear similar or derived from henna type names as for an example Alkanet it. As with myrtle, some would like to blend the histories of myrrh and henna in order to create a more mythologicaly based history for *Lawsonia spp.* as a whole.

Camphire is also a name found for henna in the Bible but as you will find in chapter 10, it has quite and unreliable past. Some modern authors have insinuated (as well as flat out preached) "camphire" is the Latin name for henna as well as a Biblical name for it too. They try to also sometimes explain camphire is an Old English version of ko'pher even. This is all quite incorrect. The Medieval Latin name for henna was "alchanna" (taken from the Persian Arabic "al-hinna") and prior to that "kypros", "cypros", "cypri" and "oleo cyprino" (meaning oil of henna), as recorded by Pliny Elder and others of the time. This is further indicated in the earliest complete Bible translation, the Latin Vulgate (405 AD) in which King Solomon's song was recorded as follows:

"botrus cypri dilectus meus mihi in vineis Engaddi"

and

"emissiones tuae paradisus malorum punicorum cum pomorum fractibus cypri cum nardo"

With the "cypri" meaning henna. Camphire is only first found in the English KJV bible translations and shows no connection to ancient names given to henna, except perhaps a corruption of "Camphor". According to some ship logs prior to the KJV Bible, camphor from the Asian *camphora* tree was already being spelled "camphire". Strangely, in the Geneva Study Bible published in 1599, all of the passages that included henna, in the Song of Songs, were missing. This included the popular 1:14 even. This shows European Christians were likely unaware of henna being mentioned in the Bible or being a Biblical plant at that time. It wasn't until the mid 1600's that investigations began to be done. The fact many were still confused (well into the 18th century) as to what "camphire" really was is supported by the words of John Wesley (1765) who wrote:

> "*Camphire*—We are not concerned to know exactly what this was; it being confessed, that it was some grateful plant."

Yule & Burnell's "Anglo-Indian Dictionary" also recounts that a mistake was made in the margins of the KJV Bible which many would turn to for explanation to foreign information:

> "It [henna/mendy] is also the camphire of the book of Canricles, where the margin of the Authorized Version (KJV) has erroneously, cypress for cyprus."

Beth, whom has been studying Latin for 11+ years and is working on her masters noted, "Camphire isn't a Latin word and doesn't really even have a Latinate form, so it is unlikely that it is derived from some lost Latin source.". It appears that other Bible translators caught this and later in the Sagond translation Camphire was changed to "troene" meaning "privet" to stand for "Egyptin Privet" which was a name used to denote henna. Most contemporary Bibles now simply use the well known word "henna". Cypri and cyprino would be true Latin names for henna. What should also be pointed out is that in the original Hebrew texts, Solomon's Song spoke of "koffer" (ko'pher) which is ancient Hebrew for henna. Later when the Latin Vulgate was created, koffer was changed to cypri. Most likely without fully investigating the background of cypri's mention in the LV Bible, some have come to the wrong conclusion King Solomon was referring to the island of Cyprus in his poetic songs. This can also be a result of the mistake Yule & Burnell alluded to. This also is noted in the classic work, the "Easton Bible Dictionary" (1897) which states:

> "The margin of the Authorized Version (KJV) of the passages above [Solomon's Songs] referred to has "or cypress" not with reference to the conifer so called, but to the circumstances that one of the most highly appreciated species of this plant [henna] grew in the island of Cyprus."

One thing that can be gained from looking at all of the various names of henna is the shear sense of the intercultural nature it has. It can not be more strongly stressed that henna is not only the domain of India or those who are of Hindu faith and the same can be said of Arabian and Muslim peoples. There is strong evidence that henna use on a large scale originated in Egypt and people of India were far from the first to use it. In reality, Egyptians, Jews, Persians and Turks were not only the ones to use henna extensively but also caused it to spread throughout the Middle East and beyond. It also had strong medicinal worth to the ancients which appears to be its main claim to popularity, even today.

I had to ponder for a good long time as to what to entitle this book. After finding literally hundreds of synonyms for henna, I wondered if I should include Mehndi &/or *Lawsonia* on the cover as well. I decided, as not to confuse people, that I would stay with the relatively universal "Henna". Now I wonder if that was the right decision. Gustave Flaubert wrote to a friend in the 19th century that he used "Lawsonia" instead of "henneh" (which was French for henna at that time) in his novel "Stammbo" because it was more of "le mot juste". Still, I don't think "Lawsonia's Secret History" has the same ring to it, especially since Lawsonia is a famous golf course and a type of bacteria today.

Chapter Two

❀

Pharaohs Gift?
Who Really Discovered Henna?

> "Your skin is a paradise of pomegranates, with the choicest fruits, henna plants along with spikenard plants; spikenard and saffron, cane and cinnamon, along with all sorts of trees of frankincense, myrrh and aloes, along with all the finest perfumes"
> —Song of Solomon 4:13–14

The title of this chapter slightly gives away my theory but I promise how I came to it is quite interesting! Please also mull over my botany and floristic kingdom research on henna in chapter 14 to get a better grasp of my presentation here.

Up until this point, writers have attempted to research Mehndi (henna body adornment) singularly and *then* use what they have found to explain henna use and history as a whole. This in return creates not only a distorted picture but also one that laves more questions instead of answers. The key is to look at the henna plant as a whole and then finger out to its numerous uses. Unlike today, where being wasteful is excepted, ancient people would rarely throw anything away. Likewise they would attempt to find every conceivable use for the meager items they had on hand. This is why if one is able to locate the first wide scale use of henna, one will find Mehndi (body adornment), cosmetic as well as medicinal uses as well. This is because all of henna's main uses are interconnected, with the exception being its essential oil / perfume form. The latter seemed to travel its own course through history from time to time. Although Muhammad did note that henna (including its use on the hands, nails, etc.) as a whole was considered a perfume. One has to wonder if this thinking goes further back prior to his time, which is likely.

While all of henna's uses may be interconnected, it is important to track henna separately from other forms of body art. This is partly due to its strong medicinal roots in history. The other reason, surprisingly forgotten by many, is henna is a *plant*. It is not something that could be created or invented by humans. First uses are thus subject to growth patterns of the plant itself. This is something apparently not considered by authors such as anthropologist Henry Field who published (in the 50's) a book attempting to link henna with permanent tattooing, kohl and facial markings (harquus). He explained that all of these ancient body markings were interconnected and created in ancient Mesopotamia. However he neglected to realize henna is a plant while all of the other types of body adornment mentioned could be created, by humans, from any number of materials. This is where botany comes into the picture. Many Orientalists, which Mr. Field relied on as the brunt of his henna research material, would explain henna originated in areas they were *presently* documenting. Writers in India stated henna originated in India, writers in Morocco stated it originated there, writers in Persia stated it originated there, writers in Cyprus and Crete noted it originating there, etc. A fine example of this is Ehrmann (1894) who wrote "Henna has been used since time immemorial in Arabia.". Easton's Bible Dictionary on the other hand notes, "the al-henna of the Arabs is a native of Egypt". To this day because of a lack of proper, objective study, the true origins of the henna plant are not 100% confirmed. Hopefully that will soon change. What is known is that mummies are the oldest remnants of henna (specifically that of Hierakonpolis) and in order to prove his theory, Mr. Field would have needed to dig up archeological evidence by way of preserved henna or literally stained artifacts. This of course was not done and Mr. Field ignored the lack of an Akkadian / Sumerian name or all of the archeological findings in Egypt. His book, published by the Peabody Museum, was on Southern Asian body markings so he of course was focusing on Mesopotamia for the origins of henna. That and looking at henna only in a cosmetic sense seriously undermined his theories, as it has for many. The biggest mistake he made was not objectively looking for henna notations. Instead, he incorporated recordation's of general, nondescript cosmetics and vermilion (red substances) as indication of henna use among people in the area of the Near East. A number of the allusions made by included authors, to baluster his theory, proved to be wrong or doctored when I investigated them. Fortunately his book seemed not to influence other works of history. It however shows how henna should be examined separately from other forms of body modification and very carefully. European Orientalist writers also do not make for very good resources when it comes to ancient henna history. This is seen by the confusion over henna presented in the 17th century book by John Gill. Obtaining information from the ancients themselves will present a better picture and is what I based this book on.

Some have realized the interconnectedness of henna's uses and thus attempted to find the area of the world where henna was first recorded in order to prove their theory of origins. Specifically they point to ancient tablets found in Ugarit which was in close proximity to the Biblical Cannan in the Mediterranean. The major flaw however in such thinking comes from the misdating of artifacts which they claimed to be in the area of 2000 BCE. Sound historical sources including Britannica states the epic stories in question are really dated from 1400 BCE. Perhaps the 2000 BCE is a corruption of the artifacts, consisting of tablets, actually being 2000+ years old. In any event, with that aside we can now look for the true first notations of henna in written form. Since the ancient Egyptians are some of the first people to use writing, in the form of hieroglyphics, it isn't all too surprising henna would be first recorded there. We also have to remember, henna is believed to have originated (or traveled from Ethiopia via the Nile) in the area of North Africa where the Egyptians inhabited for thousands of years. Thus, it would only make sense they would have the opportunity to begin using it first. Unlike with inventions, people can only begin to use botanicals if they are available to them. Pyramid wall etchings, specifically in the tomb of Teta / Teti in Saqqara (c. 2291 BCE) which mentions "henu" appear to be the first true mentionings of henna. What is even more interesting is the Egyptians took the time to describe what henna, which they called henu, looked like, calling it a green plant or shrub (Pry. Teta 1. 100). The pyramid of Pharaoh Teti was built by Wenis during the 6th Dynasty and is pretty much all that remains as a remembrance of the era. Another ancient Egyptian name for henna appears to have surfaced, "ankh imy", which was said to have a scent that could wake the dead. Perhaps henu was the henna plant as a whole and ankh imy was referring to the scent of flowers of henna; likened to how orange blossoms produce neroli oil / scent. A few hundred years later we have even more bona fide recurrence of henna (henu), this time in Ebers papyri which is dated as being written around 1500 BCE. In the papyri a great deal about henna is revealed, as it is deemed a medical document used by students. Some of the suggested uses for henna include as a hair growth medicine, snake bite cure and scorpion sting antidote. The interesting aspect of Ebers papyri is that archeologists believe its contents date back thousands of years (likely to the time of the female Hierakonpolis mummy), as if the writers were simply copying the text from a much older source. Perhaps in itself thousands of years old. These two henna sightings help confirm henna use in ancient Egypt.

Keeping on track of written accounts we return to the Mediterranean and the 1400 BCE Baal epics some make such a fuss over. The clay tablets (Ras Shamra texts) which are quite fragmented were found in Ugarit (Ras Shamra), which was once a prominent city in the ancient Canaan area because of dye production. Some, spurred by Barbara G. Walker, have attempted to use this to link henna to Canaanite people and goddess cult religions located there. Even going so far to say the Canaanite's called themselves "the hennaed people". To get the real history, I decided to ask an actual expert, Nicolas Wyatt who co-authored the history book on Ugarit and Canaanite studies. This is what he recounted:

> "The *only* apparent use of henna in Ugarit that springs to mind is in KTU 1.3 ii 2, where a number of people take "kpr shb' bnt" to mean 'henna of seven maids', or something similar. That sounds pretty vacuous to me. In my Rel. Texts from Ugarit I translate it thus, and as the context shows (esp. paprallelism with second line of tricolon) offers more consistency:
>
> > the perfumes of seven tamarisks,
> > the odour of coriander
> > and murex. (the last for color, not smell!)

> Del Olmo in his Mitos has perfumes for kpr, and the term can have the sense of smearing or anointing, which sounds better than taking it as 'henna'. We need a term || to rh. = 'ordour.'"

Now isn't that interesting and more legible. If one looks and compares that to some of the writings of Pliny, especially in the areas of perfumes, one would see quite a similarity. So much for the tablets of Ugarit being so indicative of henna use in the area. I have personally read the complete epic and it should be noted that a good ten lines directly above the "kpr shb' bnt" is completely missing and below it, it speaks of the skies raining oil (as in *perfume*). Also, KPR is thought to be the *root* word for the Hebrew ko'pher which in addition to meaning henna, itself can stand for a number of other things. KPR has long been associated with atonement via anointing oneself and not necessarily to indicate any *one* botanical (or oil). Why in the 1930's it was translated as henna is still foggy and apparently today seen as outdated (or should be). The tamarisk translation would make much more sense as it is obtained from a large, feathery palm tree. Canaan was known for its many types of palm trees including date and such notations are frequently made in the Bible and other writings. If KPR is not indicative of henna, it would mean that Ugarit too had no word for henna like other Akkadian languages prior to the Hebrew Ko'pher. The presence of Murex is also extremely interesting, pertinent to the area and warrants further discussion and clarification.

First, why the Mediterranean and why Canaanite people as the true origins of ritualistic henna use and body adornment? Is it because their name is thought to mean red dyed people? That's the only reason I could find, as there are no real artifacts, written texts or allusions by other cultures, that I could find, that the Canaanite people used henna extensively for body adornment, or anything else for that matter. There name also does not show any connection to henna. After doing more research, the Canaanite's were called such because 1.) they were traders of various merchandise, especially purple / red cloth and 2.) they themselves frequently wore purple / red clothing. I then though, in my naive days, maybe they dyed their clothing with henna? I later found this wasn't possible as henna does *not* dye cloths, such as silk or wool, red. It can only go to a dark reddish brown but only if a mordent of tin is used, which the ancients were not. Henna is never able to go bright red or even purple which was treasured by the people of the area. Of course after doing extensive research, the real truth may emerge. One that has nothing to do with the use of henna.

The Canaanite people (later called Phoenicians) were Semitic people believed to have been related to the Amorites. They developed a knack for sailing and maritime. This strong connection to the sea, allowed them to discover an extremely small marine gastropod mollusk "Murex" (purpura snail to the ancients) which produced a minute amount of red to purple dye called "purpur". One needs to understand that in ancient times, having brightly dyed clothing was a luxury. The most prized color was purple but the purple then and the color we know today was slightly different. A bloody, crimson color was the most valuable. This is because the ancients, which included Jews and Greeks, thought the color purple was the blood of the earth and a symbol of life. A sacred color showing a position of power. Why, in fact the ancients likened the color to "fresh blood". Due to extreme expense in gathering the small gastropods (83,880,000 mollusks were needed to make 1 pound of dyestuff), the cost for purple dye was extremely high. In today's money 1 pound of purpur would cost you a mere $10,000.00 USD. Hence why mostly nobles and high priests wore it, in addition to the now rich Canaanite's. The Murex, which was abundant off the coast of the Mediterranean, was the inhabitants bread and butter. It allowed them to import objects from Egypt, including cosmetics and jewelry which both men and women used abundantly. Canaanite's also had another connection to Egyptians, as the latter lorded over them for years and the expelled Hittites from Egypt became their new permanent neighbors. Between 1600 BCE and 1200 BCE, the Canaan region and all of its large cities thrived due to its production of purple cloth. The trade routes were placed, so silk coming from China could pass easily through Canaan where it was dyed purple and passed on to ships waiting to take it to areas such as Egypt. The Murex were found to rot extremely quickly after death, like crabs, which produced poor dye results. Hence the live little creatures needed their spiny shells to be broken prior to death and the dye sack (hypobrancial gland) removed. The ancients viewed the dye as Murex's blood and noted it needed to be fresh to work. Modern day researchers have found that the dyes main quality is an enzyme called "purpurase" which quickly begins to brake down after the Murex dies. A number of ancients noted that the dye produced was extremely durable, not even soaking the dyed cloth in bleach for days could remove the color. This tenacity allowed the dye to stick to walls of huge stone dye bath basins in the ancient Carthagenese city of Kerkouane where the red color is still apparent even until today. Added to this, mounds and mounds of crushed and buried Murex shells have also been found in areas of Haifa and Tyre (Tzur). Some of the mounds were measured at 100 yards tall and wide.

Murex however wasn't the only red—purple dye Canannite's held dear. Other dyes, from botanical sources are clearly recorded including Alkanet and Puwwa (*Rubia tinctoum*) which is also known as Madder. According to Pliny the Alkanet root was used as a morder and rubbed into cloth prior to dyeing with the Murex dye bath. Alkanet, which grows wild in the area, can create a number of colors for dyeing cloth including bright reds and lavenders depending on the measures taken. The

Madder on the other hand was used to produce imitation purpur and itself can be somewhat illusive. The bright red dye, which like Alkanet, comes from the roots of Madder (a native to the Mediterranean) is powdered for storage. The ancient name of "puwwa" has been suggested as the root / inspiration for the name Phoenician. Madder, also known as Robbia and Krapp, was an important fodder for animals in addition to mock Murex. The best was said to hale from Smyrna (western Turkey). Madder's name was changed to Alizarin which became the basis for the famous Turkey Red dye. The most interesting element of Madder is the fact it turns milk, bones and urine bright red…in a living being! Otherwise, if one was to feed Madder to lets say a cow, its milk would turn bright red and so would its urine. Think how magical that would have been to the ancients. The effects are the same on humans as well.

That brings us back to the fact the Canaanite's worshipped a great many pagan gods and goddesses, some of which appeared to have been taken from other cultures. This included the goddess of (sadistic) warfare Anath whose symbol was a star (like that of the Star of David) and some attempt to link to henna. In a portion of an epic about Baal and Anath translated in 1933, the latter is said to have used henna and rouge prior to slaughtering men and later wadding through their blood. However if the "kpr" is like that of the other passage which Nicolas Wyatt translated as tamarisk, it too would mean little. Also the passage was translated as such "…she departs her abode and adorns herself in rouge and henna [or tamarisk]", absolutely no mention is made of where Anath uses either the rouge or henna/tamarisk, such as on her hands or feet as some imply. In addition to Nicolas Wyatt neglecting to mention this passage, some other translations of the same epic portion suspiciously make no mention of the above line which leaves one to wonder if it is a corruption of the actual text. It should also be noted that the language in which the epic tablets were written in is not thought of as representative of the true

Canaanite dialect / writing. A. F. Rainey wrote, "…these tablets provide no clear rule by which to determine the language of the Canaanite's". Keeping this in mind, the connection of henna to Anath is simply not there, especially in the sense of using henna for body adornment and/or during weddings. Likewise there also is no evidence that priestesses ritually adorned themselves with henna either. The Canaanite's were a people that had sunken to the depths of taking a bottle of fine oil, pouring it over extremely hot coals, allowing the flames to burn blue and high and tossing their firstborn, live babies into the fire to appease their gods. Charred bodies and bones of babies have been found under alters and in mass graves honoring Tanit (Anath) by archeologists. This was for everyday commoners, the priestesses (AKA prostitutes) would self mutilate themselves regularly (self-laceration) and engaged in practices most would deem representative of the mentally ill. Somehow, a circle of priestesses joyfully hennaing themselves sounds a bit too mundane. Not only this but it's apparent that servants of Anath would want to use blood in honor of her. Either their own, of a sacrificed animal or, perhaps the ultimate blood. Murex. Why use comparably inexpensive henna dye, when the dye (blood to the ancients) of Murex was so far more valuable? It was also their bringer of riches and notoriety. Murex *was* used as a cosmetic by very wealthy Greek women, notably as a rouge. Madder and Alkanet are also possibilities. This would be in keeping with a passage from that same Baal epic where it explains in detail that "[She (Anath) w]ashes her hands of knights blood, Her [fi]ngers of gore of heroes.". One can't tell by a few idols (many of which are in pieces) that appear to have red hands (which I have yet to see) that Canaanite's began henna adornment traditions. Especially in light of their real use of blood and self mutilation, which was at times confined to the hands and feet. One should ask, if henna was so intrinsic to their society, why haven't the Greeks, Romans and /or Egyptians recorded it? Why does the Bible make mention of all sorts of botanicals growing there but neglects henna? They all certainly had much to say about their Murex production, purple / red cloth and maritime. Why does Gilgamesh mention cedarwood and other botanicals but not henna? It doesn't make sense because it isn't factual.

Nicolas Wyatt provides a theory into the henna use, if any, in the area of Canaan:

> "'Cf. the mixing of herbal extracts with oil, for medicinal purposes, in KTU 1.114.29-31, and tamarisk as perhaps the sources of seven oils in KTU 1.3 ii 2 (Wyatt 1998b [=Ritual texts from Ugarit, Sheffield Academic Press] 72 n. 15. 'Oil of peace (or: well-being, health)' (shmn shlm [sh=shin]) occurs a number of times: KTU 1.3 ii 19, iv 43, to be compared with the image of 'the skies raining oil, and the wadis running with honey' in 1.6 iii 6-7, 12-3; while oil is used as an offering in 1.41.20,21,45,46, = 1.87. 22, 48, 50.'
>
> I also believe it likely that the point of herbs and essences in the oil, and no doubt the selection of oils as well, was aromatherapeutic in intent: no doubt the anointing of various parts of the body, particularly the thin skin of the head, resulted in the absorption of substances which gave the kin or priest psychedelic experiences, interpreted as divine possession."

According to the writings of many noted Greeks and Romans, the most important product created from henna was its aromatic oil. John Gill (1730) writes "…an excellent oil was made out of it [henna] and of this with other things was made an ointment, which by Pliny Elder is called *the royal ointment*." Gill also writes that henna (ko'pher) was mentioned as part of the Jewish Misnah (specifically in relation to Judaea) so if henna was used in Canaan, it was most likely as an anointing oil or for fueling the all too common sacrificial alters. Pliny himself noted 60 BC Sidon was creating (or importing) Cyprinum perfume and even the words of King Solomon add to this conclusion as he speaks of henna in the context of an aromatic, not as a stain or body adornment. Another reason why priestesses would have had little need or interest for henna is the fact that hallucinogenics were extremely popular and revered. This included types of mushrooms (clear depiction's of which have been found on extremely old Near Eastern artifacts), opium and poppy flowers, which were attributed to certain goddesses. Such as Demeter. As Nicolas Wyatt pointed out, these substances were used to bring on divine possession and to *see* what ordinary people could not. In addition to that, such mind altering drugs also have a good tract record of bringing on looseness which would be very beneficial to *fertility* rites and rituals. Henna is by nature anti-fertility and a sedative so it makes little sense to use it for such purposes. In a psychedelic stupor it would be very easy to start cutting yourself, as the priestesses were know do and draw blood without feeling much pain. In fact, a Babylonian passage about "divine prostitutes" recorded they used pine seeds in rituals which Pliny explained were mushrooms or fungi. Hallucinogenic mushrooms, probably of the species *Amanita muscaria*. *Amanita* mushrooms by nature grow around *pine*, birch and fir trees and may have been considered the fruits of the tree. In addition to this, some take the Akkadian word for "red dye" (which shows *no* connection to henna) "tabarru" to be derived from the ancient Sumerian word TAB-BALI meaning "mushroom". Another Sumerian word for "red dye" "gannu" is believed to be from GAN meaning a red mushroom top. The *Amanita* or "fly-agaric" mushroom just happens to be blood red in color and holds extreme hallucinogenic qualities. This includes causing violent dancing, loud singing, talking to non-existent friends and periods of depressed rest. Poppy flowers (Akkadian "irru" [?]) are red too, with the sap being used for similar purposes. I think it's important to research in-depth about the ancient goddess cults use of red, mind altering drugs as there seems much more evidence compared to henna and it also makes more logical sense based on some of the women's practices. Some may not wish to do so however because hallucinogenic drug use, to many today, has a great stigma associated with it where as henna is seen as an exotic art-form. Hard nosed historians and archeologists still frown on the lack of evidence (by way of preserved henna in tombs / ruins) and designate other evidence non-supportive, thus making claims of any henna use in the Near East and in the Mediterranean pure hypothetical conclusions. All henna to goddess cult linkages stem from the 1980's adverts Barbara G. Walker made about an *unnamed* (I suppose because it was so old) statue artifact being found with red feet. Which she took upon herself to explain was representative of henna use among the followers of the goddesses Hacate and Anath. She was likely basing this on the artifacts being discovered in Catalhoyuk, believed to be the oldest urban find dating from the Neolithic period. Likely she was using photographs found in books on the subject. I asked the team working on sight at Catalhoyuk (Catalhoyuk Research Project headed by Cambridge) if any red footed artifacts or anything indicative of henna use had been found. Not surprisingly they said "no". A search of TAY (The Archaeological Settlements of Turkey) also pulled up no henna related finds at any of the digs. Likewise this would ruin J. M. Allegro's theories that henna was named after the red fly-agaric mushroom. In a very round about way, because he was unable to find Akkadian names for henna, Allegro attempted to make a strange linkage between the Arabic "al hinna" and ancient Greek words meaning cinnamon (*kinnamon*). While the writings of J. M. Allegro would technically bolster goddess cult usage's of henna, close inspection of what he was able to find and present actually accomplished the direct opposite. He was only able to locate the ancient Greek, Hebrew, Aramaic and Arabic names for henna. Usually apt at finding the linguistics of botanicals in the Near East, this further shows that it's likely people in the area didn't even use henna in any capacity until either the Egyptians or Jews introduced it much later. Anyone with an archeological or anthropological background probably are pondering why I would even include *anything* from the above two authors. The reason I do so is to show the roots of a great deal of the myths being circulated about henna's history today. Everything being spoken and written about henna has a root somewhere and no matter how outlandish, I would like to present these roots and theories in the most exhaustive manner I can. This I hope will facilitate better reasoning on the subject and perhaps new discoveries. Thus, Mr. Field's theory should also be mentioned here I suppose. He believed that the use of henna to stain the body started in ancient (or Neolithic) Mesopotamia. Remembering the information regarding the lack of linguistical evidence and the lack of real evidence at Catalhoyuk, he obviously ignored the finding that the Sumerian's in particular were not found to employ henna but instead yellow ocher (Sumerian "sa" [?]) as their most treasured cosmetic. It was used for rouge and other applications. Most researchers simply assumed Mesopotamia was using henna because the Egyptians were and there was trade between the areas for quite some time. One strange, alleged notation of henna by the Sumerian's comes from a poem on migraines. Rosemary Dudly recounted the poem as having the line, "Like a stalk of henna it slitteth his sinews.". This poem was said to be found on a clay tablet from 3000 BCE Sumer. When I searched for the poem, I found it noted other places but with no mention of henna. I also saw it being used in publications from the 1960's, so it is very possible it comes from translations done at the turn of the century or earlier. Especially when you consider all of the very Shakespearean sounding, translative words including "standeth, feedeth, knoweth", etc. Therefore it is likely incorrect &/or outdated like the 1930's Ugarit translations. The numerous college professor's and linguists who searched for the Sumerian word(s) for henna turned up absolutely nothing, so I would certainly like to see the actual word used in that particular clay tablet.

As a final note, I would like make mention that it seems beyond coincidence that the ancient peoples constantly perpetuated as first using henna or marking themselves with it, are all viewed as being white. This includes the ancient Moroccan Berber's, Sumerian's, Canaanite's, Anatolian's, etc. India is included, but conveniently only after the Aryan invasion. This leads me and hopefully you as well, to wonder if something consisting of bias lies at the heart of some 19th and 20th century research. If true or founded it would explain the apparent lack of research done in Egypt, North East Africa and Jewish areas and the glossing over of henna related finds in these same locals, as they are not considered "white" by many. Hopefully this book will be one that is seen as being unbiased and driven by the goal only to document the glorious past of henna, no matter where it takes us.

EGYPT, EGYPT, EGYPT

The climate in Egypt, while dry, always seems to stay 65 Fahrenheit, even in winter and has the Nile to provide marsh like wetlands which henna loves. Many archeologists believe henna to be a common bush to grow along with roses near the houses of ancient Egyptian estates. According to both Pliny and Dioscorides, the best henna and Cyprinum by far was produced in Egypt near the city of Canapus in the Delta. The ancient city of Canopus was located in North Egypt and known far and wide as a pleasure city for the wealthy (likened to Monaco of today). So it isn't surprising the best henna was said to hale from there. Another indication of mass growing of henna comes from the Egyptians themselves whom speak of a mysterious "Land of Henu" which was supposedly placed on or around where the Biblical land of Goshen was.

A district of Egypt, Land of Henu which may be translated as Land of Henna was said according to pyramid texts to have fields of henna and a great estate for pharaoh Rameses II (1225 BCE or older according to the Bible). In Genesis 47: 11, Goshen was described as being "the very best of land" in a pastoral sense at least and it's also interesting to see in 46: 4, where it speaks of Jacob's family having to travel there because of severe famine in the land of Canaan. In deed the land is quite well irrigated in the area as the Nile sprouts out into a number of separate channels which forms a kind of oasis which reaches the sea. The more modern name of the area, "Saft el Henna" (synonyms being Saft el hinna and Saft el henneh) appears to draw further linkage to the ancient use of the land. Egypt also replanted henna all about Karnak and the temple of Amun to recreate the dynasty of Ramses the I which they claim extensively used *Lawsonia*. In fact Ramses I's mummified body is believed to have just been identified by the Michael C. Carlos Museum. Like his son and grandson the mummy has very unique features such as a bridged nose and is being examined for traces of henna. The Egyptians, whom never seemed to speak of having to *import* henna from other areas, appear to have had an extremely long span of time to experiment with the botanical. They also had the know-how to produce wet-chemistry and advanced medical techniques (compared to other cultures anyway). This would have allowed them to easily discover henna's cooling, UV protecting, antifungal, etc., qualities prior to other peoples, including body adornment as well. Holding a ball of henna, which was mixed with liquid to form a clay like consistency, to cool the body is quite an ancient invention. These balls would intern stain the hands yellow to orange depending on the length of time they were held or rolled by the feet. These marks that could be left behind just incidentally make the majority of ancient depiction's of stained hands, palms and feet. Henna *was* an ancient Egyptian remedy for heat prostration and headaches and was mentioned in an epic flavored tomb inscription about pharaoh attempting to use it to heal himself where it states, "then his majesty reached the fields of henna...". It goes on to say how he was still not healed of excessive heat in the limbs and had to resort to priestly magic tricks. Another possible origin of henna staining, is that from using henna in the hair. The Egyptians were known far and wide for their elaborate wigs and hair-weaves, many times stained with henna. Wig shops have been located and so has henna hidden in the wigs. Henna is also believed to have been used on ones natural hair as well and from extremely early on, pharaohs dyed their beard with henna. While the Egyptians seemed ingenious when it came to cosmetics, they certainly didn't have plastic gloves. Anyone who colored their hair as much as they did with henna had to notice stained nails and hands. Especially on the nails, it only takes a few minutes for the color to take. With all of the other cosmetics the Egyptians mass produced, including kohl, it isn't too much of a stretch to see them harnessing the cosmetic staining power of henna as well. Even to create designs on the skin. Archeologists have found examples of tattooing on the skin of some mummies, where the color was injected under the skin. Pictorial examples of such include a plate from 1400 BCE of a lute player with a Bes insignia on her left thigh. Some tattoos were simple, primitive dotted and dashed designs. Likewise henna has been found on the feet and hands of mummies. While henna stains on the hands as a result of practical uses may sound mundane, it would explain why decorating the hands transcends so many cultures and religions. Why many women see it as being simply cosmetic, like kohl for example. Just like henna, kohl was (and still is in some areas) thought of as being both medicinal and cosmetic. This shaped its real use throughout the ages. Likewise it's very possible henna staining stems from comparable medicinal use which later became outdated.

In addition to leaving extremely old writing, the Egyptians left something even more valuable. Themselves. Preserved human tissue is worth a great deal more than paintings in some remote cave. Through DNA and other scientific studies, researchers can learn a lot including what the mummy died from, how old he or she was and even traces of botanicals are left on or in the body. This is how it is known Egyptian people have been using henna, at least on their hair since 3400 BCE. In fact some of the oldest mummies ever found, including a woman from Hierakonpolis (Burial 16), had traces of henna in her hair weave. Specifically *Lawsonia inermis*. This is said to be the *oldest* example of hair dye use in all of Egypt and possibly the oldest use of false hair (wig) anywhere. Some say the nails of some mummies are also stained with henna but coloring can also be caused by simple oxidation…so studies must be done on a case by case bases to determine if it is in fact henna preparation. Some point to the Bulaq 3 papyrus (Cairo Museum) and the n. 5158 papyrus (Louve), which are from the Roman period, as making mention of henna being used on the nails. 11th dynasty mummies so far have been the oldest found with hennaed nails. Likewise it was said, at least by Victorian era authors, that henna was used to stain the cloth used to wrap the mummies. Ramses II is stated to have such henna stained wrappings. Prior to the invention (or mass use) of mummification, archeologists have found what appears to be examples of Egyptians preserving the skeleton and hair (head) only. Boiling off the dead ones flesh and then wrapping the cleaned bones in cloth. In these cases evidence of henna used by that person would have been destroyed and can explain why ever earlier forensic examples of henna use haven't been located. Many museums and places housing mummies are likewise reluctant to do testing in fear of disturbing or harming the diseased person. Very few mummies are in fact ever unwrapped today, unless they are already damaged. Hopefully with the advent of more and more less invasive diagnostics, this will change.

More evidence that Egyptians were not simply using the flowers of henna is the finding of preserved henna leaves (*Lawsonia inermis*) in not only Late Kindom tombs (c. 600 BCE) but also Ptolemaic tombs in Egypt as well (c. 200 AD). Such direct artifacts anywhere else have *not* been found. Some have stated the tomb of King Ashurbanipal of Assyria had not only a cosmetic box filled with henna powder but also the kings body itself had indication of henna staining. I found the latter at least very peculiar because another royal tomb only had skeletal remains. I also was unable to find any information pertaining to the tomb excavation. I asked James A. Armstrong, a Mesopotamian Archaeologist at the Harvard Semitic Museum about King Ashurbanipal. He explained the following:

> "...7th century King Ashurbanipal's tomb has never been found. I have checked to see if henna has been reported in any of the few known Assyrian royal tombs, but it never has, as far as I can determine. Neither have any cosmetic boxes been identified in those tombs. So, the short answer to your question is that such information is not correct."

This shows that henna was dear enough to the Egyptians yet not in other cultures, that it was thought proper to place it in the tomb prior to closing it. Perhaps for use as medicine or in the hair during the after life. The Egyptians would place everything they cherished, as if they were going on a long trip, in their tombs. All of this hard evidence supports the theory of henna use stating in Egypt and spreading as trade routes to and from the area opened. Unfortunately, many archeologists gravitated to the flashy gold and bejeweled items, in the numerous tombs discovered, neglecting to document possible botanicals left preserved inside. This was the case with various wet-chemistry formulations which after over a hundred years of sitting on a dusty museum shelf are just now being analyzed. It's quite probable that henna was left in much older tombs but perhaps was overlooked, ruined by grave robbers or as an oil form, evaporated. In Nubia, a popular folkart item made is a henna container called a "gofad". I asked the Oriental Institute, who has Nubian gofad's on display if ancients ones like it have been found. Ms. Emily Teeter who is in change of the artifacts replied that none had been in fact found yet. Perhaps Egyptologists can use the Nubian folk-artifacts as a model in order to look for more ancient examples in Egypt.

The writings of ancient Greeks and Romans also shows how henna was used and exported from Egypt. According to Theophrastus (c. 285 BCE), one of the first botanists, the perfume of henna, "Cyprinum" was a costly scent much like "The Egyptian". Such perfumes as "The Egyptian" were being sold at 400+ denarii a pound. Compared to the 1 denarii a day which was a normal mans working wages, one can see how only the elite could afford such a perfume. Today, that would be about the equivalent of $20,000.00 a pound! Of course most people probably didn't buy a pound at a time but still one can gain a sense of worth the henna perfume and henna itself had. Just because one lived with perhaps an abundance of fresh aromatics growing about them, didn't mean they had precious perfumes and oils as well, especially if they were of a lower class. Food, housing and clothing would have been more of a priority and most people probably didn't have the equipment needed to do their own extraction's. The time it took to grow the henna or wildcraft it, harvest and turn it into perfume, placing it on great ships, etc., all added expense and more importantly exoticism to Cyprinum, which the rich craved. According to Dioscorides, "henna was not found growing everywhere", thus it had to be carried to other countries and regions. He also said the henna unguent could keep for up to 3 years. Returning to Theophrastus, he wrote Cyprinum should be clear in color like all of the other fine perfumes of his time. Pliny also recounts that by his time, a few hundred years after Theophrastus, henna alone was used as a commodity and simply called "Cypros" or "Appelle cyprus". He stated…"A tree found in Egypt is the Cypros, which has the leaves of the ziziphus". The ziziphus is a type of bush that grows in arid areas of the Middle East and India; the leaves of the two do resemble each other depending on species, some of which produce edible fruit. Pliny also took much time to explain about henna being from Egypt, as if he were saying it was indigenous to the area. The seeds / berries of the henna were soaked, boiled and pounded in fine oil, presumably in Egypt and sold for about 5 denarii. Also one can see the price plummeted. Instead of equivalent of $20,000.00 a few hundred years ago, it appears henna is worth about $250.00 a pound during the era of Pliny. Perhaphs this is due to parts of the Mediterranean and Cyprus now producing henna oil as well. That of course was just simply henna seed oil, the full Cyprinum was still being produced as well as a few other perfumes containing henna. Dioscorides, whom lived and wrote about the same time as Pliny, noted that Egypt too produced the best henna and it was extremely fragrant. Yet he also said it was not abundant everywhere in the Old World and thus needed to be carried to various areas in the form of an ointment.

Well into the Ptolemaic era, Egypt continued to corner the market in perfumes and fragrant herbs. They also stated to be known for Alchemy. This is shown by a papyri dating from around 250 AD called the Leyden and Stockholm. Actually the Leyden is one copy and the Stockholm is its twin. Both appear to, like the Ebers, have been copied from older sources. This is noted by some of the information which resembles that of Pliny and Dioscorides. These papyri's were hidden in the sarcophaguses of their owners. Due to a decree which stated all Alchemy documents were to be burned and destroyed, they are quite valuable for their contents. In the papyri there are hundreds of recipes including how to make fake Murex, which the Egyptians called "purple snail", in addition to simulated gold and pearls. One of the more interesting recipes (to us anyway) is that for making heliotrope colored jewels for jewelry. A hard resin like substance, which were the nodes of the "tabasios" (Indian Banboo tree) was prepared and then soaked in a dye bath of henna mixed with wood tar. The outcome would be pretty violet looking jewels. They and all sorts of other fake precious materials became so popular that Egypt became frightened that their wealth would become undermined and ordered the secret recipes destroyed at once. Who knows how many other henna recipes were destroyed as well.

OTHER ANCIENT MENTIONINGS OF HENNA

Archeologists have apparently discovered yet another name for henna, this time the Greek word "e-ti". This e-ti apparently found on Mycene clay tablets which date from the 13th century BCE and include recipes for various types of perfumes. Since some say fresh henna only grew on the island of Cyprus, some think the recipes were used for dyeing cloth instead with henna leaves. Of course if one follows the recipes of the Roman Pliny Elder whom explains henna seeds / berries (sic) were used to create a perfume, perhaps they were used instead of fresh flowers in the Mycene recipes. In India henna seed oil is created for medicinal use so the possibility is there. Likewise maybe oil of henna flowers was mixed with other ingredients. Dioscorides explained that henna ointment could keep for 3 years and certainly could be exported from Cyprus or Egypt within such spans of time. Unfortunately not much is left from the ancient city of Mycene as it was totally destroyed by fire. All that has remained is the clay tablets written in a hard to decipher dialect. Some figures have also been found with red markings on their palms but one can not really tell from that if people in the area really used patterned henna body art.

The Bible provides a source for our next mention of henna. As part of poetic songs by King Solomon, henna is mentioned twice in the book of Canticles. Written in Jerusalem around 1020 BCE, the poems (found at the beginning of this chapter and chapter 3 of this book) seem to be more speaking in the realm of henna's highly aphrodisiac scent. During Biblical times it appears some of the best henna was produced by Jews. According to Dioscorides, the second best henna grew in Ascalon in Judaea. Pliny Elder agreed, saying Ascalon did produce the second best. This is quite interesting however. Around the Dead Sea, in Biblical times, especially around King Solomon was nothing of how it was today; a salt encrusted waste land. The area was bustling agriculturally and home to the famous En-gu'di gardens and vineyards of King Solomon where henna was grown. Unfortunately things seemed to take a dramatic turn for the worst in the 5th to 6th century BCE. The gardens and the peoples seemed to simply disappear from the landscape. Geologists, measuring the contents of the Dead Seas floor found indication of severe settlement changes. Changes that may have come from major geological problems including earthquakes. This could explain why suddenly En-gu'di no longer was a source of henna. Ascalon on the other hand was, like Sidon, an ancient seaport and also has gardens which produced various produce including henna at least around 60 AD. Returning to the mention of henna in the Bible, Hebrew for henna being "ko'pher", it is only found in three places in the entire book. It is also one of only three flowers mentioned in the entire Bible. Although the actual reference to henna itself is found only three places, the actual word "ko'pher" is found in a number of other areas of the Hebrew Scriptures for a total of 15 times. They were however translated to mean other things; ransom (8), bribe (2), satisfaction (2), pitch (1) and sum of money (1). Like we today have a color *orange* and a fruit called *orange*, using the same word to mean a number of other items is not uncommon in ancient times. Ko'pher in the sense of henna is known however by comparing various translations of the Bible including the Latin Vulgate which mentions henna as Cypri, otherwise henna. The Jews are said to have been encouraged to keep small gardens while enslaved in Egypt which included medicinal and culinary plants including henna. The henna, most likely in oil form, was said to have been used in baths to ward off leprosy and the dye, which is brown, to color wool clothing / fabric. If the Biblical land of Goshen and Henu, where the Israelites were enslaved for over 200 hundred years, were one in the same, it would explain why henna was grown later in En-gu'di and other Jewish areas abundantly. Once again, at that time henna was an expensive commodity and was reserved for special occasions or treatment of leprosy in the case of poorer peoples.

The Island of Cyprus is a fine example of the Greeks knowledge and ancient use of henna. The Island has actually had a number of names including Makaria meaning blessed, Aeria, Aspelia, Ofiousa, Amathusia and the ancient Hebrew Kittim. Believed to have been inhabited since 3000 BCE, the Greeks influenced the island around 1500 BCE. It was in the time of the Greeks and some think specifically Homer that Cyprus is to have gotten its present name. Meaning "henna plant", Homer called Cyprus "kupros" in his epic Iliad (850 BCE) because so many henna plants flushed on the island. The name stuck! Cyprus was long an island of great wealth and the inhabitants enjoyed all sorts of luxuries including cosmetics and perfumes. During at least Pliny Elder and Dioscorides time, 60 AD, Cyprus was said to make the third best Cyprinum and also produced the 3rd best quality henna around. This is most likely due to the Greeks constant borrowing from the Egyptian culture, especially in the area of perfumes. The Maycene tablets would also explain a long use of henna in perfumes by the Greeks as well. As you will read a bit later on, the perfumed oils and ointments were also important to Greek herbal medicine. The word Cypros however should not be confused with *Cypris* which means "from Cyprus". This separate meaning but like sounding and spelled word may have led to some mistakenly associating henna to the goddess of love Aphrodite (Venus) who was at times referred to as Cypris. Although myrtle was attributed to her, henna shows no real connection to the goddess. Cypris gave forth to Cyprian which means "belonging to cyprus" in addition to denoting a lewd (harlot) woman. Most likely the last mention was referring to the many priestesses which inhabited the island.

Moving on, next we have henna mentioned by the first botanist, Theophrastus who lived in the 3rd century BCE. In his day, henna oil and such was held on the same level of other perfumes such as "The Egyptian" and could easily be sold for 400 danarii (which is the equivalent of $20,000.00 USD today) a pound. He notes in his book "On Odors" that the Cyprinum created from henna should be clear in color. What is interesting is Cyprinum predates Cleopatra. Even though it is frequently associated with her, it most likely was an extremely old concoction that the Egyptians were using and exporting for some time. It also gives an interesting glimpse into the worth of growing henna and how it could be quite profitable and lucrative. The "worth" of henna seems to be quite a pattern, with it the constant use by the rich and royal.

Its quite apparent that the Greeks and the Hellenistic era had a profound effect on the spread of henna. Since the Greeks has a pretty long tract record of using henna themselves in a number of ways, they encouraged others to do so directly such as in medicine or indirectly such as in the Iliad, through the emergence of writing. Not just any writing but that of influential people whom had their works copied and distributed to anyone who could read. A fine example of this is a fort and castle seemingly named henna. The Hellenistic Fort Cypros from around c. 300 BCE was located near where the wall of Jericho was thought to have once stood. While it isn't noted if henna was grown there then, later when King Herod (65 BCE) took it upon himself to rebuild Jerico there was henna in addition to many other plants including myrrh and palm trees in the great gardens of his palaces. Still protected by the ancient Fort Cypros, Herod was said to have made "vast reservoirs, a hippodrome and amphitheater". Also interesting is King Herod's sons wife, an Arabian, was too named Cypros. Some attempted to change the Queen's name to Cypris and translate it as "Venus" but today historians pay more close attention to Flavius Josephus (60 AD) who noted Cypros as her name. Since Josephus was writing around the same time as Pliny, who also uses "cypros" to describe henna, it seems pretty probable he was in fact referring to henna. This allusion to henna is all that is made by Josephus in his writings.

HENNA USE IN C.E. AND A.D.

We owe a good deal about henna's use throughout history from the words of two men, Dioscorides and Pliny Elder who wrote c. 60 AD. According to Pliny, henna was known for its odor, having a sort of sweet scent and the best grew in Egypt. He also likened the leaves to other shrubs including the ziziphus and christthorn. He called henna Cypros and in its finished perfume form, Cyprinum. The henna itself, called Cypros or "appelle cyprus", not surprisingly appears to have been a commodity separate from Cyprinum and costing about 5 denarii a pound. That would be the equivalent of $250.00 USD a pound today. While modest compared to other aromatics including myrrh and cinnamon, it was still expensive enough to be considered a luxury item. The normal mans day wage was around $50.00 USD so its obvious henna wasn't being used by everyone, especially the poor. According to Dioscorides, henna colored the hair orange and it was red like the blood of a dragon, so apparently it was being used in that capacity as well. Also what is interesting is Cyprinum was thought of as a *mans* perfume, light and delicate. Women on the other hand enjoyed more pungent, spicy scents. A reversal of what women and men enjoy or think proper today. However, with henna essential oil's unique attributions (see the chapter on henna's aphrodisiac side), I wonder about this notation. Since Dioscorides was a physician, he of course noted henna's medicinal applications which included healing fractures and nerves. This is extremely important because the writings of Dioscroides was translated into many languages including Arabic and was carried throughout the Old World, especially the Middle East. In an 11th century Arabic translation, a picture of henna is found. Simple looking, the artist recorded it as being a short shrub with green leaves and light brown bark. While stylized, it looks much like henna in a small bush form, which is especially seen with henna grown for its leaves. These illustrations were many times copied from earlier works of De Materia Medica (c. 60 AD). The extreme popularity of Dioscorides work most certainly helped to spread the use of henna further throughout the world. Galen (180 AD), a Greek doctor who lived in Rome also had much to say about henna, which is included in chapter 5 of this book. Like Dioscorides, Galen's writings were translated to many languages and remained the basis for medicine until the 16th century.

Many classical authors were recorded as making mention of henna, when in reality they did not. A number were simply speaking of general, nondescript cosmetics or herbs which were liberally translated by Orientalist writers as "henna". This is seen with such authors as Ovid and Juvenal. Another of these authors may be Arnobius (311 AD) who was from Sicca Africa and wrote 7 books on Christianity. One of the English translated passages reads, "Was it for this He sent souls, that, forgetting their importance and dignity as divine, they should…seek for cosmetics to deck their bodies, darken their eyes with henna, curl their hair…". He then goes on about men using these items to look feminine and how harlot women used such ornaments. Certainly, it would seem Arnobius was bashing henna use, especially for Christians, as it was the pagans he was associating its use with. The darkening of the eyes with henna however didn't sound quite right. Although Discorides did make mention about henna being used medicinally in or around the eyes, I still decided to compare the English with the Latin version (from the 16th century). Specifically book 2, paragraph 41. Much to my *dismay*, I could find no word denoting henna in the books original Latin translation. If henna really was mentioned, it would be safe to think his words prevented a number of Christians from using henna. Cosmetically at least. However, this may simply be another case of "implant henna" for color. As seen by Ms. Edward's translative actions concerning an ancient Egyptian love poem. Thus, it would not help us in understanding henna's use among the people of the area.

China emerges as a country to use henna quite late compared to other locals but can help show time spans it took for henna use to spread. While there may be mention of henna in Qi Han's 4th century AD work "Nanfang ts'ao-mu chuang", otherwise "Plants of the Southern Regions", it is a work of agriculture predominately and doesn't do much to show actual use by the people. Henna has been known to grow wild in Southern China and while not native, that could have come as a result of trade or Indian / Persian influence. The ancient Pharmacopoeia's of China do *much* more to shed light on the unique ways the Chinese would incorporate henna into their lives. Specifically henna's status as a "Cao Yao" herb / blend shows that it is mentioned in even older works than Qi Han's works including "Shan Hai Ching of Medicine" from 250 BCE and also the "I Ching". It also means henna appears in later compilations which include information from those previous works. We have to also remember that earlier notations of henna use in China may have been destroyed, as when China was unified, the emperor ordered all history books to be burned and annihilated. Even so, it may be safe to say henna appeared in China around the Han dynasty and specifically when Wu-di pushed the Turkish speaking Xiongnu out. This opened up the "Silk Road" for China and allowed Wu-di to send exploratory armies to India and Persia to bring back all sorts of items for the royalty. Cao Yao means "Secret (domestic) Family Remedy" and would indicate henna's main use was on the skin, in sometimes a cosmetic manner. During the Tang period (618—907 AD), henna was said to be sold in exotic bazaars in China by merchants from other locals. The Tang period is also likely when henna (Zhi-jia-hua) was introduced to Japan (along with tea) by the Chinese as it is called fingernail flower (tsume hana) there as well. Tang poets scathingly wrote about Chinese women using foreign cosmetics, which was highly looked down upon. As anything foreign was abdicated as being barbaric or horrid. This is what likely stopped the use of henna as body art in China from ever taking hold. That and a ban of tattooing. The work "Ben Cao Kang (Gang) Mu" (Great Herbal) by Li Shih completed back in 1590 AD has by far the most content about the medicinal and main uses of henna in China. The book was so large and so in-depth that it took Li Shih over 30 years to complete (and I thought 3 years to write this book was a long time). The majority of the information on the Chinese use of henna can be found in chapters 2 and 5 of this book. Henna is also considered a "Guan Yao", meaning it is an official Chinese remedy of all the ancient Pharmacopoeia's including the 11th century "Han Ching". This is like a "status" code the Chinese smartly created to allow one to know if a herb or remedy was used in various times without ever having to read the actual book. Single herbs were many times not as popular as giving patients a complex mixture of herbs as prescribed. These were then named after their main ingredient or what they were used for. Henna too had such herbal formula titles, "Wu Bai Zu Hai Na" and "Yuan Wu Bai Hai Na", with the "Hai Na" denoting henna.

Since the Hai Na sounds and even looks much like henna, perhaps this shows origins of its true introduction to the Orient. In China, perhaps because of the proliferation of plant pests henna was not immune to, *Lawsonia inermis* plants actually create galls. This is really the only place I have seen such occurrences. The gall grows on the henna's leaves and is scraped off for use in both skin and hair treatments. All this being so, henna was however *not* widely used in China. It was strictly reserved for the higher echelons of women. This could have come as a result of henna being scarce in China. It has been recorded that finding a henna plant is like finding a centuries old ginseng root! What may make it appear China used henna abundantly is the fact they had their own Lawsone producing plant called "Garden Balsam". Few plants contain Lawsone but the Chinese Garden Balsam and its very close relative "Jewel Weed" do. Being that this is so, henna was simply a reinvention of the wheel of sorts for the Chinese. It however may have had more Lawsone and melanin stimulating properties than their native plants. Women would use henna to turn their skin further yellow in color in addition to keeping the skin UV protected from the sun. Its apparent in many cases where a country has their own native, comparable plant that they will many times use the latter instead of the alien as price is usually less high. Today, henna is found growing in the Yunnan, Fujian, Guangdong, Guangxi and Jiangsu providence's of China.

In a 2nd century BCE work appears to be the first notation of henna in Indian texts. The word used was "Madyantik" with the "a" missing. It has been assumed to be Madyantika which is considered representative of henna in ancient Sanskrit. Notations of henna are also found in a complex recipe for a black hair colorant (for the recipe read chapter 4) which is in keeping with the Karma Sutra (320 or 540 AD) texts which too provides a hair coloring and conditioner recipe containing henna. Due to numerous allusions to other yellow and red botanicals being used both religiously and cosmetically in India, especially turmeric, saffron, barleria, myrtle flowers, etc., and later henna being likened to these other plants, it appears henna's use in India was an introduced practice &/or plant. A plant that became more romanticized and misconstrued in India after the independence in the 1950's. These early Sanskrit texts would seem to dismiss theories henna was brought to India along with Persian horses in the 8th century AD. Numerous Indian scholars repeat that henna came to India as a gift from Egypt. Since there has been notation of Egyptian ships reaching India, including that of Strabo (20 BCE) who noted "With the advent of the Ptolemies, Egyptian trade with India began and was developed by the Romans who succeeded them...". He also recounted that when he visited Syene he saw about 120 ships sailing from the Red Sea port of Myros-Hormos to India. Henna coming straight from Egypt could have been a possibility...it still would mean it was an introduced product and it could have taken many years for seeds to come in order to be grown there. Since India had many *other* colored cosmetics, there wouldn't have been much of a rush. There also appears to be a connection between when China and India have their first written notations of henna. For the Chinese it was the Han dynasty and for India, it would be the Mauryan dynasty. Specifically henna could have been introduced when Askora sent missionaries to Libya, Egypt, Syria and Macedonia. It is likely henna reached India first and then China obtained it from there. These missionaries likely brought back very stale henna or only the botanical and no explanation about how to use it. Some point to the Ajanta caves as proof that Indian people were using henna on the skin prior to the Mogul invasions of the 16th century AD. I believe, after studying numerous pictures that they prove otherwise however. The Ajanta caves (1st century BCE through 6th century AD, although the dates are fiercely being debated) of India are thought to be some of the oldest examples of Indian art around. The caves which are located in the state of Maharashtra (where Bombay is found), are actually man made and housed Buddhist monks who practiced ceremonies and also lived there. After around the 7th century AD the caves were abandoned and the forest overgrew the doors so they remained hidden all the way until the 19th century. Unfortunately, unlike Egyptian tombs which are located in extremely dry surroundings, the Ajanta caves have fallen to water damage, mold and fungus, mineral deposits, cooking fire / incense silt, etc. Paying special attention to caves 7 and 11, I saw no evidence of henna. Especially on a Queen sitting on a pillow, her hands and feet clearly outlined and colored yellowish, the same as the rest of her body and face. In another, what looks like a King with wives all about him, no coloring appears on any of their hands either. However, in some areas there appeared to be extreme water damage. Some portions of the paintings were completely missing &/or the backing of the wall peeking through the paint. In other areas, at times, the feet from far away looked darker than the body but upon closer inspection, there were large water stains which extended past the feet and discolored the whole area. Restoration has attempted to be done and may have led to denaturing of the art to reflect henna use as well. In keeping with other countries such as Malaysia who specifically point to Islam as the bringer of henna body art traditions, I think its safe to say so too was India. Nevertheless, I believe the use of henna actually on the hands started prior to the finale Mogul invasion by a few years. Many forget Tamerlane's invasion of India in the 12th century which could have initially led to henna use on the hands, feet, etc. The Ajanta caves however do little to prove anything about henna's use in India. The art, which is clearly Buddhist inspired, is located in caves in Maharashtra which begins India's Florida like nature out into the Arabian Sea. This causes one to ask why this remote area would have the first evidence of henna use? Also, since henna isn't intrinsic to Buddhisem, especially as body adornment, why would pious monks be using it in depiction's and offerings? Another cave setting called Ellora has also been suggested as having depiction's of henna on the hands. If its anything like Ajanta, I wouldn't be all to surprised if that proves false as well. Ellora is however considered much younger than Ajanta.

Perhaps because art is subject to heavy personal interpretation, another attempt to make henna an integral, ancient part of Indian culture and society is the claim a beloved Hindu sage named Agastya wrote about henna in his Vedic texts known as "Agastijamamouni". He was supposedly quoted as saying henna would dispel diseases, which sounds plausible, and then notes henna is "the guardian of the pulse" or something similar. Said to have lived anywhere from 4000 to 5000 years ago by people of India, Agastya was recounted as having been born (like a tadpole) in a water filled vase maintained by nymphs and when he was an old man, would become enlightened and write in Tamil on palm leaves, various verses. He also was a devoted worshiper of Ganesha, the ancient Hindu Elephant God. That last sentence really appealed to me that this *ancient* henna connection was untrue. As explained in detail in chapter 6, an integral form of worshipping Genesha was with the use of turmeric. Turmeric is a yellow to orange spice which, when in a paste form, can be used to dye the skin a number of colors including varying shades of yellow, orange and deep red. Known as Haridra and a few other Sanskrit names, turmeric is frequently found in Vedic texts. Since the spice is believed to have originated in India, this is not surprising. The problem has come when some have tried to claim a word, "Mehaghni" said to mean turmeric in ancient Sanskrit, is the root word of "Mehndi". Thus some say that this "Mehaghni" is found in ancient Vedic texts, perhaps that of Agastya and proves henna has been used as a yellow dye since ancient times in India. Firstly, I personally have never seen mehaghni, as it is spelled, used to denote either henna or turmeric in Sanskrit or any language. It appears to be a corruption of Mehndi itself. Secondly, if mehaghni means turmeric, is found in ancient Vedic texts and is the *root* word of Mehndi (or Mehndi is derived from this word)...how does this denote henna use in ancient India? Wouldn't that mean turmeric was being used and then when henna was *introduced* was named after turmeric? Thirdly, Mehndi really means "Myrtle" in Sanskrit, only later in Hindi did it begin to be used for henna as well. Of course the above, hard to follow explanation some Indian scholars give for their bases of claiming henna is found in ancient Vedic texts is rarely presented to you as I have done. What you are likely to see is something to the effect of "henna is found in some of the oldest Vedic texts", and that's it. That's why it's important to fully investigate claims which have vague references before jumping forward and stating them as fact. I believe the Indian Lexicon when it notes the 200 BCE book "Navanitaka" and the word "Madayantika" is the oldest mention and name of henna in Sanskrit. The mehaghni claim on the other hand I believe is incorrect. Mind you I have recounted it just as it was explained to me and it really makes little sense. In any event, returning to the supposed notation by Agastya about henna, it's very likely he was not talking about henna but instead turmeric. Agastya wrote in Tamil, a language with *no* similarity to Sanskrit. He also was said to write on palm leaves that a certain Indian cast maintains. Thus, such writings would have had to be recopied by hand and then distributed. Hindu leaders also have written that people should avoid secondary Vedic texts because they preach hate, religious intolerance and violence in the midst of fighting between India and Pakistan. This really makes it difficult to research henna use in India using modern texts. The blatant intolerance ruins logical thinking in many circumstances and henna has been sadly placed in the middle of a cultural tug of war. Even worse, this has effected the translation of ancient texts and caused many problems with dating of artifacts. Being that this is so, I have attempted to use truly old translations of Indian texts and also fully investigate claims of ancient henna notations. Returning to Agastya just one more time, his "the guardian of the pulse" in regards to henna (or really turmeric) is interesting. Turmeric just incidentally is used in Ayurveda in a number of pulse related health conditions, whereas henna is *not*. Turmeric is indicated for poor circulation, clearing the Chakras (which can be considered like pulse points) and cleansing the channels of the body. This and Agastya's association with Ganesha worship really undermines the probability of any of his works being indicative of henna use in ancient India. For more on this you may wish to read chapters 6 and 9. The "politically correct" conclusion is that henna appears to have reached India around the Mauryan dynasty, for use by the rich and royal as hair color and perfume. The "Night of the Henna" and patterned henna designs (perhaps even henna designs in general) on women came with the Mogul invasions and slowly caught on by the 16th century.

After the 11th and 12th centuries, the notations about henna becomes more apparent and frequent, as seen in other chapters of this book. Based on ancient texts I have produced this conservative timeline. If you will note, it features religious groups as well as ethnic groups. Any area that was heavily influenced by such groups most likely too would have used henna, at least medicinally....unless there was great hatred for the newcomers. While the cosmetic use of henna could have come and gone with the whim of women, the medicinal aspect would not and have been the driving force behind henna use as a whole.

Conservative Henna Timeline:

—Egyptians c. 3500 BCE (or likely even earlier)

—Jews c. 1725 BCE (or later according to pyramid inscriptions)

—Greeks c. 1300 BCE (or earlier according to some professors)

—Arabian tribes / traders c. 900 BCE (or slightly earlier)

—Persians c. 850 BCE [?]

—Africans (Nubians) 800 BCE [?]

—Chinese c. 250 BCE (or slightly later)

—Romans c. 200 BCE

—Indians (India) c. 200 BCE

—Sicilians c. 50 AD

—Berbers c. 500 AD [?] (Likely earlier but with no written history, impossible to pinpoint)

—Turks c. 600 AD (Likely earlier)

—Islamic's c. 600 AD

—Indonesians c. 700 AD

—Africans (Moors) c. 700 AD

—Sri Lankans c. 700 AD

—Japanese c. 700 AD [?]

—Spaniards c. 712 AD

—Malaysians c. 1400 AD

—Gypsies (Romany) date depends on local. This would give more credence to the notation that henna (body art) traditions did not take hold of India until around the 12th or 15th century. Gypsies are thought to have emerged from India around 1100 BCE and in 200 AD reached Persia. They only seem to have henna traditions from Spain and depending on where they were settled, as opposed to being the ones to spread henna further throughout the Old World.

*The "[?]" denotes that written sources are not available but keeping with historical events, trade, use of henna later on, etc., an approximate date may be given.

Main Factors of My Conclusion of Henna Use Beginning in Egypt:

—Preserved Tissue / Hair / Nails—Mummies have been found with henna traces on their hair, nails, hands, feet and possibly the gauze used to wrap them.

—Oldest Name—Henu spelled in Hieroglyphics is found not just once, in one place but continuously throughout Egyptian history.

—Direct Artifacts—Preserved henna leaves have been found in tombs from the Late Kingdom onward.

—Origination—Many works site North Africa as the origins of the henna plant. Its optimal growing conditions mimics the climate of Egypt and the banks of the Nile. This could be due to henna originating in Ethiopia and traveling to Egypt (naturally) up the Nile.

—Pictorial—Tomb paintings of women with orange stained hands.

—Major Exporter—Well into the Victorian era, Egypt and Nubia remained the major exporter of henna. Even today they are one of the top 5.

No where else in the Old World could I find all of the above. I believe that even the hardest nosed historian and archeologist would find this (and information I provide in the chapter on henna's botany) persuasive as a basses for where henna use originated and from which it spread. As to how henna was actually *discovered*, which is a popular question, I surmise, from my experiments with henna, that Egyptians noted dried and dead henna leaves falling into the Nile (or a tributary) which turned the water a rich rust / red color. Such happenings have been seen in streams even today where henna grows close to the water and the leaves fall in. They then experimented with the dried leaves and found they could dye hair and skin with it. They could have been drawn to henna in the first place because of its inviting flowers and then discovered a red dye could be created from them. Henna leaves are not a one step, "pick the leaves and you have red dye" deal. The green leaves, when fresh are unable to dye above a light yellow. Especially skin and water. Both dried and fresh leaves impart an olive green color to oil and wax. So a measure of work was needed to use henna as a dye and from outward appearances, one would be hard pressed to imagine it could dye orange and so forth. As Sir Thomas Brown noted in his "strange but true" notes, "Alcanna being green will suddenly infect the nails and other parts with a durable red.". I find henna to be very similar to tea. Even though the tea tree (*Thea sinensis spp.*) is thought to have grown in Prehistoric China, it wasn't grown commercially until the 8th century AD. So, simply because a botanical is growing in an area doesn't mean that people are using it. That's why the finding of the pre-dynastic mummies and hennaed hair is so important and telling about the true origins of henna's use.

Another thing to think about is how henna would hold up on trade routes. Remember the words of Dioscorides, henna isn't found everywhere, thus it must be physically taken to other locals. If henna was somewhat scarce in his time, it was likely even more scarce hundreds and thousands of years before. Henna oil evaporates rather fast, leaving only an unusable sticky film on the container it was carried in. That would be of little use to people. The ointment could last, according to Dioscorides, for up to 3 years so that certainly sounds more travel friendly. Henna leaves can be stored for about 1 year and still produce good color results *but* only if they are not allowed to absorb any water. Moisture of any type ruins the dyeing ability so the leaves would have had to be stored in very dry conditions and away from sunlight which causes premature oxidation. Then there is the powder, which apparently was carried in much the same manner as salt. Sacks that is, usually of jute. From my own experience, henna powder only lasts a bit over 6 months. It was likely stored for 6 months before I obtained it, so lets say 1 year. This is however based on henna kept in an opaque, air tight container in a cool, dark environment. It is likely the containers used to carry henna in ancient times were a lot less protective. Thus, likely shaving a month off of the powders longevity. From Egypt or any of the major henna growing points, one could use the above observations to create a radius of how far it was likely henna could really travel. Based on how fast ships and camel / horse caravans could move from area to area. Here are some sobering pre-UPS travel statistics. A horse can travel 5.0 km per hour. A camel, 6.5 km an hour. A fully packed caravan could cover about 20 to 30 miles a *day*. A ship could cover about 10 km an hour if there was a good wind. According to some c. 1600's logs, it took Armenian merchants starting in the Mediterranean a whole 45 to 70 days to reach the Persian Gulf. Using my Reader's Digest World Atlas, I measured Greece was 900 km's from the Delta of Egypt. Lets estimate a ship would be covering 5 km's 24 hours a day, leaving from Egypt. It would take about 9 and half days to reach Greece, which is actually not bad. It is easy to see why the Delta area was a popular, ancient grower of henna. On land however things get much slower. It is around 800 miles (on a map that is, curvy roads and inclines make the distance more I'm sure) from the Delta of Egypt to Baghdad. At 20 miles a day, it would be approximately 40 days for the caravan to meet their destination. It is likely people living very far from major growing areas and oasis received very poor quality henna, imparting only a yellow tinge. This is why seeds were likely preferred so they could grow it themselves or people would buy the all too popular henna ointment. In our fast paced would of Air Mail and next day delivery, many fail to consider the limitations of henna long ago and its ability to hold up. It is very possible henna in an ointment form was its exclusive mode of long distance travel. This may mean henna body art was severely restricted to areas that either grew henna or were in close proximally to regions that did.

MURKY HENNA TRADITIONS AND MYTHS

The various sections of henna's history which prove untrue are many times directly related to the following myths which I will decipher here.

Many have a strong conviction that the Berber's of Morocco were some of the first people to use henna in the form of body adornment (staining), specifically in patterns. Since henna is believed to have grown very early on in North Africa, this isn't so far fetched (although the conditions of Morocco are not of the tropical nature henna desires). The problem comes with documentation. The Berbers up until today, had no actual written language of their own. Their only means of recounting their history came from folktales, songs, etc., which over time can become highly denatured from the original. The same has become of the Egyptians myths which are not relied on anymore for historical worth. This makes trying to date their use of henna almost impossible and purely speculative. What is known is the ancient Egyptians actually referred to a people living in Morocco in addition to the Romans and other cultures quite early on. However their use of henna was not mentioned, leaving some to wonder if they even were using it in ancient times. If a peoples excelled at something, they would many times be associated with it. Having fancy designs on the body certainly would have been something noticed by the ancients. The Berbers are and always were either nomadic or an agricultural people. Today many Berbers have intermarried with the other ethnic groups that have moved into the area. The majority also converted from Christianity to (Sunni) Islam in the 7th century AD which some point to as the true starting of henna traditions in the area. The fact the Berber word for henna is directly derived from its Arabic name also does not help the situation. Another story being circulated, perhaps by Moroccans themselves, is that Cleopatra's daughter Cleopatra Selene (meaning moon) married Juba II in Morocco and had a traditional "Night of Henna" there prior to her wedding. Anyone with a history / archeological background would of course find this far fetched. Juba was from Africa where henna was not an integral part of the culture yet and Cleopatra Selene was said to have been married around 25 BCE. She was given the ancient areas of Mauretania and Numidia which would be between modern day Morocco and Tunisia. Very little is known about her and there certainly is no explicit details about her wedding or usage of henna. Some have even pondered if she ever existed.

Next, the subject of "the Hand of Fatima" warrants further investigation and discussion. Many perceive the hand talisman frequently seen in the Middle East and the double thumbed hand of India as being a Mehndied or hennaed hand. While perhaps examples seen today are created to look like a hennaed hand, its first uses appear to have nothing to do with henna or body art. I believe from research I will put forth here, the hand talismans greatly predates henna's deliberate application to the hands and is the reason henna today is used to divert the "Evil Eye". Not visa versa. Thus, I haven't used hand talisman artifacts to document the history of henna's use. There simply isn't any reason to do so because the two are unrelated. Firstly, the Hand of Fatima, which is also known as the Hasma Hand in Arabic and the Hemesh Hand in Hebrew appears to have been used since at least 800 BCE. Both Hemsa and Hemesh means the number 5, in relation to the extremities on the hand. Prior to this, the two symbols used on necklaces (the hand and crescent moon) appears separate. In ancient Semitic settlements, the crescent moon was used to represent the moon deity which had a number of names including Sin. This is seen in on numerous artifacts including those of Assyrian and Babylonian art. The hand talisman seems to be a result of worshipers of the moon raising their hands in prayer and admiration. This is seen on ancient Canannite artifacts unearthed which depict two hands stretched before a downward crescent moon above. On extremely old hand talismans, there is absolutely no ornamentation. It is simply made of a smooth material. Above it, downward crescent moons are hung which is attached to a beaded necklace. Later, on more fancy hands, an eye was placed in the middle of the palm or the folds of the skin would be depicted. At times special sayings (prayers) would be inscribed on the hand in Hebrew or Arabic which could serve to confuse some into thinking it was artistic henna designs. Both Hebrew and Arabic calligraphy can appear very pretty and artistic. When the Romans and other ethnic groups began using the hands, they many times decorated them with figurative motifs and made the pinkie and 3rd finger bent down to the palm. This intern made the talisman into the "Hand of Power". The Evil Eye and the hand talisman are interconnected as the crescent moon shape representing the moon deity. Most likely because of the moons effect on the tides and such, the moon goddess was in charge of water. Water to the ancients equaled life while things that were dead, many times dried up to dust. This sort of thinking has led many researchers and historians to believe such beliefs are Semitic &/or started in areas that were arid and desert like, such as the Middle East and specifically ancient Sumar. Remember henna has no ancient Akkadian names and appears not to have been used in the areas of Sumar, Assyria or Babylon in ancient times. From the moon goddess beliefs arose the Evil Eye superstitions which many still see as a real sickness. The curse occurs from a glance that causes a living being to dry up. Thus, the most thought at risk for the Evil Eye were nursing mothers, nursing animals, fruit trees, vegetables and so forth. A person whom casts the Evil Eye isn't necessarily evil themselves. It's something that is believed to be far from ones control. Hence why kohl us used around the eyes as it is said to prevent one from casting (projecting) the Evil Eye. The hand, perhaps representing secret prayers to the moon goddess for protection, was worn around the neck and fastened to other places to divert the Evil Eye. It isn't until Islam that a connection to henna seems to first appear. This is most likely do to henna first starting to be used for Evil Eye protection in that era. Islam forbids anyone from worshipping the moon, sun or stars. Turning the hand talisman into a hennaed hand of Fatima would have allowed for such *side* worship to continue in secret. The same would have been true for Judaism.

So comes in the "Evil Eye" itself. In numerous cases around the 18th century onward it is noted that women wear henna to divert the *Evil Eye*. Since henna is cooling and moist when held in the hands, it's not all too surprising. However, henna being used in ancient times in regards to the Evil Eye isn't noted like kohl is. Contrary to popular beliefs, the Evil Eye isn't as wide spread as many think in the Old World either. Areas such as Korea, China, Burma, Taiwan, Indonesia, Sumatra, Thailand, Vietnam, Cambodia, Laos, Australia (aborigines), New Zealand (aborigines) and Africa south of the Sahara do not have traditional Evil Eye suppositions. Therefore anyone in these countries (or migrating from them) would not have had Evil Eye protection as their basis for using henna, including in a wedding setting. Unless introduced by Muslims. Likewise the main form of protection from the

Evil Eye was the hand talismans, lewd hand gestures, eye amulets, blue beads (symbolizing water), crescent moon shapes, etc. If one fell sick from the Evil Eye, eggs and holy water is used as a cure. Henna appears to be a much later addition to the superstitions in the Middle East and India and may have come as a result of attempting to hide worship of the moon deity in the 7th century AD onward.

Chapter Three

❀

Cleopatras Liquid Seduction
The Aromatic Side of Henna

"As a cluster of henna my dear one is to me, among the vineyards of En-ge'di."

—Solomon's Songs of Songs Canticles 1:14

Henna is quite a versatile bush indeed. In addition to its cosmetic and medicinal qualities, essential oil of henna was extracted from the petals, which are considered some of the most fragrant flowers on earth. According to both Pliny Elder and Dioscorides, henna was "famous for its odor" and had a "very pleasant scent". Buddhist monks are said to also use henna flowers in their ceremonies and women would use fresh flowers in their hair for personal perfume. Cyprinum was the name given to the sweet smelling perfume by the Greek's and was said to be Cleopatra the VII's (69 BCE) most favorite perfume. It should be noted that Cyprinum and oleo cyprino are indicative of henna perfume / oil but Cyprium which sounds and *looks* very much like it is not. Without that simple "*n*", the meaning is changed to "aes Cyprium", otherwise "metal of Cyprus" which was later shortened to one word (Cyprium) and corrupted to "cuprum". Our modern copper is a result of the Greek Cyprium / cuprum. Cyprus was the only producer of copper (Cyprium) for the Roman armies hence the metal being named after the island. Also there is a flower that grows abundantly on Cyprus with the partial Latin name Cyprium which too would not be indicative of henna or its use.

It was recorded that Cleopatra had all of her Nile barge sales soaked in henna perfume (Cyprinum) and wore it when she met Mark Antony. Perfume in that time had more meaning and was an important aspect of daily life. It was seen as a form of healing and protection in addition to being aromatically pleasing and a mood enhancement.

Everything was pure and real, chemically produced scents were not in existence and therefore many perfumes created did have medicinal qualities. The Egyptians have always been thought of as the first to create large quantities of perfumes and essential oils for both personal and temple use. Aromatics were also frequently burned in the form of incense. Perfume was used to create perfection spiritually instead of on a physical plane according to Egyptian thought. It's actually a vast understatement to say the Egyptians took their perfume making very seriously. They even went so far as to appoint deities to each of the 700 scents they held dear and Ra and Nefertum over all perfumes. To them, perfumery was medicine. One 2000 / 2600 BCE papyrus from the reign of Khufu speaks of "fine oils" and lists aromatics that would make "every god gladdened". Some of the more well known oils to be used included cedarwood (*Cedrus atlantica*) and myrrh (*Commiphora myrrha*) which actually had to be imported to Egypt from Lebanon and the Red Sea areas. Being that henna is thought native to Egypt's arid area (or perhaps Ethiopia) and the fragrance of the flowers so blatant, especially at night, it would be hard to believe Egyptians were not also using it as well. If they had the capabilities of importing aromatics from other areas as early as 3000 BCE, why not utilize something growing right in their own backyard? These oils were also formulated into healing creams and salves (ointments) for the skin and henna was found to be an ingredient in some of these ancient recipes, one being the famous and very sacred "kyphi". Kyphi was used both aromatically and as internal medicine by the Egyptians. The name kyphi is actually the Greek form of the ancient Egyptian original hieroglyphic pronunciation "kapet". The Greek essayist Plutarch (100 AD) wrote that Kyphi was for ones who liked "those things which delight most in the night." and also that kyphi "lulled one to sleep and brightened the dreams." There have been a number of authentic recipes found including the following:

Ancient Egyptian Kyphi / Kapet

Take equal parts of Juniper (*Juniperus communis*), Acacia (*Acacia spp.*), and Henna (*Lawsonia inermis*) and allow it to soak in fine oil or wine. At this same time, raisins (or grapes) should be allowed to soak in wine as well. Both for the span of one week. Then gather together equal parts of Cardamom (*Elettaria cardamomum*), Calamus (*Acorus calamus var. angustatus*), Cinnamon (*Cinnamomum verum*), Peppermint (*Mentha spp.*), Bay laurel (*Laurus nobilis*), Galangal (*Alpinia galanga*) and Orris (*Iris germanica var florentina*) root. Combine and grind into a fine powder. Add to that one tablespoon of honey and one tablespoon of Myrrh (*Commiphora myrrha*). Drain the herbal mixture and raisins from the oil or wine and mix them into the powdered herbs and honey. Add enough wine used to steep the herbs with Turpintine (*Pistacia terebinthus*) and raisins until a thick paste forms. This can be used as a salve or dried in bricks for incense use today.

D. M. Stoddart wrote that Cleopatra likely burned kyphi in large alters on her Nile ships / barges. Thus, getting a double dose of henna. Stoddart also explained another recipe for kyphi which included the roots of *Andropogon* and *Acorus* were mixed with the oils of henna, cassia, cinnamon, peppermint, *Convolvulus*, pistacia, juniper, acacia and cypress…which was then allowed to macerate in fine wine. This mixture was added to bees honey, myrrh and resins, dried and then burned. Many times flowers were simply placed in fats or olive oil to extract their scent and later used for anointment and religious rights. Some of the more fragrant essential oils lingered in the tombs and could still be identified when Egyptologists reopened them for the first time in thousands of years. In addition to this, special preparation rooms have been found for the creation of perfumes in ancient temples, one in particular being Denderah. These rooms were the domain of healers / priests, the precursor to the medical physician of the time. Discovered on the walls of these preparation rooms were actual formulation notes used by the healers carved into the stone. These rooms are said to date from about 4500 BCE. Further indication that henna was most likely first used in the form of essential oil by the Egyptians. Like henna leaves, the dark red to brown oil from henna flowers too can stain the nails or add color to the skin. The first recognized medicinal physician in history was Imhotep of Egypt. In addition to being thought of as a healer, he was also a magician. In that time, before the revaluation of Greek physicians such as Hipprocrates, it was common for a physician to be a magician, healer and priest all in one. This would account for the simultaneous use of perfumes in temple and personal applications, such as incense and oils. At that time, there was little distinction made between cosmetic, perfume, edible, ritual, medicinal and so on use of essential oils and botanicals. Everything was intertwined. Everything was connected. The kohl used around the eye would have been just as medicinal and sacred as the Cyprinum applied to the skin.

Cyprinum, while sometimes referred to as a single "note" (or oil) perfume, was actually a blend of a number of botanicals. This would make sense because henna flowers do not produce much oil and adding other botanicals would help stretch the batches created. This would also explain why Cyprinum was described as being clear or light green as opposed to its real color of red / brown when in absolute form. A number of recipes for Cyprinum have surfaced including this one:

Cyprinum Perfume

Cardamon (*Elettaria cardamomum*), Henna (*Lawsonia inermis*), Calamus (*Acorus calamus var. angustatus*), Wormwood (*Artemisia absinthium*—wormwood oil is today considered poisonous by aromatherapists and thus recommended not to be used today!) and Rosewood (*Aniba rosaeidora*).

One of the earliest recipes is from Theophrastus (300 BCE) whom explains the main ingredients are henna flowers, green olive oil (most likely used as a base or carrier oil), asplathos and cardamom. If you will note Theopharastus writes about Cyprinum 300 years before the era of Cleopatra, showing that henna perfumes were in use prior to her birth. In a recipe given by Pliny Elder (60 AD), henna seeds, green olive oil, sweet flag, myrrh, cardamom and southernwood are combined to create the infamous scent. Pliny, who seemed to be slightly confused about henna to begin with probably got a very un-cyprinum smelling perfume as the seeds have their own acid based perfume and flavor components. Dioscorides also provides recipes which calls for henna, green olive oil, rainwater, aspalathos, sweet flag, cardamom, myrrh, cinnamon and old wine. These recipes give a glimpse at ancient Egyptian perfume making as the above 3 men, especially Discorides studied much about their medicine and perfume making. Many of the papyri they transcribed has since disappeared. According to Dioscorides recipe, a bit of oil was combined with water and the herbs were placed in it to macerate, except for the myrrh and sweet flag which stepped in wine. These two are then combined and boiled. The plant material is strained off, leaving an oil and then the *fresh* henna flowers are added. This is allowed to macerate and then strained off. To create a stronger scent, the step of adding henna flowers could be done another one or two times using the same perfumed oil. Dioscorides also adds that one may add cinnamon to the blend as well. Much debate has come over Pliny's recipe which calls for henna berries / seeds. Many believe his information seems to be from a *different* source. In any event, his recipe may be for areas that did not have access to fresh henna flowers, further indicated by the word "E-ti" found on ancient Greek Mycene tables. E-ti is thought to mean henna and since henna is thought to have not grown on the Greek mainland and fresh flowers could not make the trip from Cyprus, dried seeds / berries had to be used instead. Although the fresh flowers may not have been able to have been used in certain locals, the oil extracted from the flowers could have been. This oil could then be used in place of fresh flowers as an additive to various perfumes. Henna absolute is said to be quite volatile and evaporate rather quickly in extreme heat. It is unlikely however the ancients were able to extract absolute but instead used the "soak the flowers in olive oil" method which would create a longer lasting perfumed oil. Dioscorides noted that such a henna concoction (ointment) could then last for 3 years. It must be noted that the berries / seeds of henna do in fact produce an oil which is at times utilized in India and Ayurvedic medicine; perhaps giving credence to Pliny's recipe where oil extracted from the henna seeds was used as a base or carrier for the perfume. Cyprinum wasn't the only perfume to utilize henna, another which Pliny calls "The Royal" is supposed to incorporate only the finest and most exotic scents known to man. While not a real recipe but more of a simple list, the ingredients include henna, cinnamon, cardamom, spikenard, sweet flag, jujube, wild grapes and a few other spices. More evidence that henna was thought of as an important scent suitable for royalty and the well-to-do. In the time of Pliny, henna was about 5 danarii a pound to purchase, with a normal days wages being about 1 danarii. The best was said to be produced in Egypt and on the island of Cyprus. Especially during the Late Kingdom era, Egypt was known far and wide for creating some of the best perfumes and essential oils. This led to most people simply spending the extra money to have it exported from there as opposed to trying to create it fresh in their country of origin; where perhaps henna did not grow in abundance. While the finished perfume was said to be clear in color like other expensive perfumes, the addition of olive oil could explain Pliny's description of it being light green and some even think henna leaves were being used as a green colorant. Some have likewise mistakenly attributed Cyprinum to be comprised of Cyprus Reed, or the oil extracted from it. This most likely came as a misinterpretation of the plant Pliny calls "Cyperos". That single "e" causes the meaning to no longer denote henna. This is shown by Pliny taking the time to describe this Cyperos as having dark olive green leaves which "resemble leeks" and extremely fragrant roots. The best he said came from the temple of Hammon in the Libyan desert. Since Cyprus Reed's flowers were not used in perfumery and either the seeds, it's quite doubtful Pliny or Dioscorides were using it in true Cyprinum. Especially when Pliny and Dioscorides took the time to describe the henna (Cyprus) plant in detail. Dioscorides wrote the henna plant had white, fragrant flowers, full branches of olive "like" leaves except they are larger and softer to the touch and black seeds. Pliny wrote that henna had very fragrant white flowers, etc. Confusion over Cyprus Reed being the basis of Cyprinum was most likely fueled by 18th century speculation about what "ko'pher" actually was in the Songs of Solomon. A number of Bible historians attempted to implant Cyprus Reed in henna's mention. Another theory, presented by Mr. Henry Field, is that what Pliny and Dioscorides talked about was henna (Cyprus) only used for perfume and that *other* (mysteriously unknown) names were used for henna that made body art. This likely stems from his lack of finding evidence of documented henna *body art* in the ancient Near East. Some of the names proposed included the Greek kino and kene and the Hebrew KNA. None of these showed any connection to henna however. Greek linguists told me they could find no link between the 2 words mentioned. I found that the ancient Greek "lugx kene" translated literally as "empty heaving". Kene was also a corruption of the Turkish word kIna as well and may be the true root of the Greek Kene to henna connection today. Kino was also discounted in the first chapter of this book because it is the red juice or sap from botanicals not related to henna. The KNA alone only had a strange association to the word "God" in Hebrew texts or in conjunction with other words, to signify other people in the Hebrew Bible translations. Linguists too failed to find a KNA connection to henna. When looked at as a whole, Pliny and Dioscorides were talking about the henna plant in general and what it was *really* popularly used for in their times. If henna really had all sorts of popular names, the above two men likely would have listed them. Dioscorides listed the many other names of myrtle. Pliny did say that "ligustrum" was what henna was called in Italy but he was likely confusing the two plants. Still, if there were many other names used in the Near East and Mediterranean, they likely would have mentioned them. Mr. Field obviously ignored Dioscorides notes about henna (Cyprus) dyeing things orange and red. Also he didn't take into consideration Galen, Avicenna and others who continued the documentation of henna used both in ancient times and their respective times. Thus making linkages to henna as a whole.

Egyptian women, especially musicians and dancers, were thought to also wear pillars of thick fat on their heads called Bitcones which had been impregnated with aromatics. As the heat from the head escaped, the fat would melt and the aroma would cascade down their hair, waxing it along with their bodies. Murals show this practice being kept alive well into the 15th century BCE and recipes have also been found. Due to the fact clothing and cosmetics trends and usage changed little over the years of the Egyptian empires, perfume used by Cleopatra most likely was also used by other Queens such as Nefertiti (1372 BCE) in much the same manner. One thing to be noted about Cleopatra VII is she did have an affinity for henna, in addition to Cyprinum and coloring her hair with it, she was said to use another concoction called Aegyptium which combined henna with fine oil, honey, cinnamon (*Cinnamomum zeylanicum*) and neroli (*Citrus aurantium var. amara*) which is extracted from orange blossoms. Archeologists believe they have also recently located her perfume creating laboratories not far from the ancient En-gu'di in an area called En Boquet near the Dead Sea. However, if the perfumery is from Cleopatra's era, 30 BCE, it most likely was not used for Cyprinum of henna as severe geological changes hundreds of years prior left the land useless for major agriculture. Instead it was likely used for Cleopatra's other love, balsam.

THE SPREAD OF PERFUMERY

> "The leaf of the henna plant resembles that of the myrtle. The blossom has a powerful fragrance; it grows like a feather about 18 inches long, forming a cluster of small yellow flowers."
>
> —Baker Nile Tribes, 19th century

As trade routs become better, people started to highly value the various aromatics being imported to their area. So much so, that it became its own currency and even, at times, more precious than gold and jewels. With the monumental exodus of Jewish peoples to Israel from Egypt around 1240 BCE (or earlier according to the Bible) so went many of the gums, essential oils and botanicals the Egyptians held dear. Due to the Jews being slaves of the Egyptians, they received a first hand knowledge of aromatic use and healing and may have been the second ethnic group to use henna regularly in essential oil form. Not only this but the biblical land of Goshen where the Israelites were held in slavery was later discovered and thought to be a large henna growing and producing area of Egypt. Referred to by Pharaoh as the city of Henu (henna) in pyramid hieroglyphic etchings. This would make a nice linkage of henna use to Jewish peoples. By the time of King Solomon and well afterwards, the Jews were known for producing some of the best henna flowers and perfume, especially in the area of the Dead Sea. These perfumed oils were also used to create ointments and by the time of Nehemiah, both skilled women and men ointment mixers were allowed to join a guild for their field. Fragrant ointments and oils were extremely important to the ancient Jewish peoples, perhaps even more so than to the Egyptians. Pronounced makeup was the domain of the harlot and women were taught that inner beauty was more valuable. These women still needed the protection from the harsh elements that makeup afforded but without the vibrant colors. Hence the use of heavily perfumed ointments and oils which, if henna was used, would have provided sun blocking qualities and conditioning to the skin. To keep it soft and non-chaffing. It would also act as a deodorant and of course was thought to have medicinal properties. To the Jews, ointments and oils were their cosmetics. It was also used for anointing the dead body, most likely for antibacterial and deodorant purposes. It would have been a final show of respect, as many times only the most expensive oils were used. At $20,000.00 USD in today's money, a pound or more, henna certainly would have been one of those *expensive* oils! Around the 5th or 6th century BCE however, archeologists have found evidence in the Dead Sea which indicates major geological changes occurred including earthquakes. This caused flourishing agricultural areas known for henna oil production such as En-gu'di to become wastelands and the peoples populating the area to appear to vanish. This may have caused a shifting of henna perfume production to Sidon and perhaps even (Herod's) Jericho. John Lightfoot (1650) in his book (From the Talmud and Hebraica) quotes a passage about Jericho's gardens as follows:

> "The place also feeds bees and produceth opobalsamum, cyprinum and myrobalanum; so that one might not call it amiss, "a divine country." "

Unfortunately, the spread of the Egyptians perfume usage also led to its "watering down" and less emphasis being placed on the healing aspect of the essential oils and more on their aromatic qualities. In 525 BCE the Persians conquered Egypt which is quite interesting and may have led to their introduction to henna. Later, the use of aromatics began to make its way throughout the Mediterranean, especially Greece and Rome. The Greeks admitted to learning a great deal about perfumery from the Egyptian Suna's (physicians) and Herodotus in 425 BCE began to document it more in detail after visiting Egypt. He wrote that the Egyptians would kill a cow and stuff its carcass with honey, bread, figs, myrrh, frankincense and a whole host of other aromatics, bathe it in oil and toss it onto the alter flames. The Greeks incidentally were the main ones attempting to extract secrets of perfumery from not just the Egyptians but the Persians and other local cultures. They were especially interested in the Egyptians treasured "kyphi" but not for burning at night for a restful sleep. Instead they wanted to use it as a love potion of sorts. When Cleopatra died an untimely death, the Romans took over Egypt (30 BCE) and matters did improve slightly. Although the Romans began using essential oils much as the Greeks did, but in more excess, they were forced to return more to Egyptian thinking about healing with perfume. Mass plagues broke out over the vast territories the Romans now owned. Sick and dying people were of no use to the Roman empire therefore something had to be done. Unfortunately, the Romans used essential oils so recklessly and in such excess that it really caused them to go bankrupt and led them to ruin. There is truly such a thing as "too much of a good thing". Not only that but the respect for essential oil and perfume was also gone, replaced by the thinking that perfume is for pleasure, not for healing.

INDIA'S FAMOUS HINA PERFUME

Atar or Attar, meaning *smoke, odor, wind* and *essence* is derived from the Arabic word "*itr*", and is a popular method of essential oil extraction in India. Originally invented by Muslims, most likely Persians, the still used was called an Alembic, it was later adopted by India and other parts of the Middle East as a viable way of extracting the "scent" and "essence" of a botanical. This is believed to have occurred around 1600 AD in India although some claim significantly older perfume creating equipment has been now located. While it isn't certain if aromatics were what was referred to in Vedic texts which marked a start of the "Science of Life"; they were highly prized by the people of India and most likely started to be produced on a large scale after Mogul occupation in the 16th century AD. Attars differ from essential oils in that instead of being pure and undiluted, they are rendered in a base of low grade sandalwood oil. Although the sandalwood is quite pungent itself, due to high pressure distillation process called "hydro-distillation", its scent is displaced and absorbs the scent of the botanical rendered in it. Therefore Attars are categorize more in the realm of perfumes than pure essential oil, as all are adulterated with sandalwood and at times, mixtures of other cheaper essential oils.

The henna essential oil created is traditionally known by the names "Hina Attar" and "Mandee Attar" or simply Mehndi perfume and is mainly produced in April and May (although it is also made in the fall months in some areas) when the flowers are in bloom. Like Cyprinum, Hina perfume in India is a blend of many aromatics, especially spices which include nutmeg (*Myristica fragrans*), cloves (*Syzygium aromaticum*), ambrette seeds (*Pimpinella anisum*), cardamom (*Elettaria cardamomum*) and turmeric (*Curcuma longa*) in addition to the henna absolute, jasmine and rose oil, spikenard (*Nardostachys jatamansi*), oakmoss (*Evernia prunastri*), juniper (*Juniperus communis*), sandalwood (*Santalum album*), valerian (*Valeriana fauriei*), lurel (*Laurus nobilis*) and the like are blended together and has been traditionally produced in Uttar Pradesh. There is no real single Hina perfume blend, as a number of families produce it and have strived to keep the exact ingredients secret from each other and the world. The numerous other ingredients being added to the henna oil, which is extracted separately, is probably to stretch the perfume as a whole. The single note or pure henna oil was traditionally called "Gulhinna" and is stilled produced today by many henna growers for extra income. This involves soaking the crushed henna flowers in oil, sometimes for up to 20 days. The longer the process, the more expensive the henna perfume. We know henna was extremely expensive to buy in oil form in India because it was reserved for temple and court use and being given as a gift to royalty. This is echoed by the Sus'ruta Samhita (400 AD [?]) which explains that henna (Madyantika) was used in expensive ointments including one called "Angarage" which was reserved for royal use. Angarage also contained sandalwood and was used as an all over body perfume and offering to Krishna. People of India would wear perfumes according to what season it was as well. Jasmine was traditionally used in Spring while henna was used in Winter. Since Spring is a very important fertility time of year, it sort of casts light on how henna was really viewed. Most commoner people probably had little if any access to the henna in Attar or perfume form due to the expense it took to extract it.

Hina Attar (or essential oil) can be likened to the scent of Boronia (*Boronia megastigma*), in itself prohibitively expensive, which has a warm, woody-sweet fragrance with rich floral undertones. Henna's pungent nature is said to be a result of it containing b-ionone and various esters and aldehydes. Henna essential oil has also been likened to Chloranthu's (*Choloranthus spicata*) and Lilacs (*Syringa vulgaris*) scent. Due to this, we can see why it was blended with sandalwood (*Santalum album*) and rose as it forms a perfect, aromatic synergy. Its scent is also described as being phenolic or bitter and having a tenacious, leafy backnote or otherwise tealike. The powdered leaves of the henna plant are very green and grassy smelling and may explain these undertones the essential oil carries. *Personally* I find henna's scent to be absolutely intoxicating. Drawing on my background of Aromatherapy and numerous essential oils in my collection, I venture to say henna is somewhere between jasmine and a very subdued ylang ylang. Very sweet but with an exotic twang. It's simply unique though, really there is no comparison that can be made I think. The flowers do not readily yield their essential oil, hence causing the need to use large quantities of plant material to get a small bit of oil. Likened to that of the Rose (*Rosa spp.*). This in turn causes the price of the oil and its value to be quite high. It was just as expensive and labor intensive in Cleopatra's time to produce as it is now because henna does not respond to the newer essential oil extraction methods. Imagine the gallons that were needed to soak the sails of her barges, what expense that must have been. Henna essential oil was definitely a luxury of the rich and higher classes. The resulting henna essential oil is anywhere from a dark orange to dark brownish yellow to reddish color and is quite viscous such as Benzoin (*Styrax benzion*) is. Many find that the essential oil, in its concentrated form is too strange to ever wear. The scent wafting from the flowers on the henna bush is found by many extremely inviting. Thus many think henna's scent is best enjoyed straight from the flower. William Loring wrote that fresh henna flowers were carried by Egyptian women to perfume themselves and uplift them, noting they were the women's constant companions. I find it really amazing how one small cluster of flowers can be so aromatic. It literally lights up an entire room. African women (especially Muslims) create a fresh potpourri like perfume called "Kikubu". Its main ingredients are jasmine flowers, henna flowers, rose petals, pompia leaves, kilua flowers and mkadi flowers. Kikubu is also dried and mixed with rose water and olive oil to form a perfume especially used by brides.

In Ayurveda, which is discussed in much more detail in chapter 5, essential oils were important in treating the Chakra's. According to contemporary Ayurvedic texts, henna essential oil was used to open and balance the Chrown Chakra "Sahasrara" which is associated with the pineal gland. Before their association with certain organs, Chakra's were thought only as points of energy on the body and could be strengthened by the application of very specific botanicals to the point in question. A special blend was created and applied to the area:

Chrown Chakra Oil

3 teaspoons essential oil or Attar of henna
7 teaspoons of neem oil

This oil was then applied to the corresponding Chakra, this time being the Chrown Chakra, in a counter clockwise motion. Cold compresses were also created using henna / hina essential oil in the manner of 10 drops to a few cups of water for which a cloth was soaked in it and used to apply the concoction. Sometimes it was allowed to simply drip onto the body as well.

AROMATHERAPEUTIC USE OF HENNA

> *"...the Orientals excel in spices, in dyes and drugs, henna, otto and camphor..."*
> —*Ralph Waldo Emerson*

What about henna essential oil's effect on the body? Perhaps in the sense of Aromatherapy? Well, although henna essential oil has been described as having a "somewhat medicinal" scent, very little is known about the healing properties of the oil, if any exists at all. Having a wonderful scent doesn't automatically mean it can be used for Aromatherapy. Some scents that are heavenly, can only be used for perfumery. A few examples include that of Boronia which some compare henna's scent to and Gardenia (*Gardenia jasminoides*). They have reported aphrodisiac qualities but that's about it. The chemical compoition of the henna essential oil is the key to both its aroma and its biological effects. The main components in henna essential oil has been noted as being esters, aldehydes and b-ionone. Easters are typically very soothing, calming and have the ability to kill fungus. They are balancing, electrically neutral and add a fruity aroma to essential oils. Aldehydes are also calming / sedative in addition to being anti-inflammatory in nature. The b-ionone (or beta-ionone) has both an anti-fungal and bactericide quality to it. A-ionone (alpha-ionone) is also present in the essential oil which is considered as causing allergenic side effects in some. Henna essential oil is also stated as having alcohol components. These alcohol's lend extremely pleasing scent and also a deodorizing and antiseptic quality. They also mean an oil will be uplifting and usually pretty non-toxic. All in all, henna would fall under being an oxygenated compounded oil and likely has over 100 volatile constituents. In Islamic medicine, the perfume of henna has been traditionally inhaled to cure headaches caused by extreme heat which is quite interesting. In India it has been used in Ayurveda to help calm one, relieve anger and frustration. It is also noted as being calming and relaxing to the body and mind and a few drops of hina attar on bed pillows was said to bring one deep sleep. Aromatherapy sources point out it is best when used in a diffuser (such as an aromatic lamp or lamp ring) or in a bath / massage oil. You can make your own massage oil from crushed flowers and a bit of sweet almond oil (read chapter 14 for full instructions). The attar is also said to reduce pitta and kapha yet increase vata. One traditional anointing oil made in India is called Kaapu and combines henna essential oil (attar) with hibiscus flower oil. The major factor causing a lack of documentation about henna essential oil is the fact it is *extremely* hard to obtain. This can be caused by the small nature of the flowers. On one occasion Dr. Bala Subramaniam M. told me of when he ventured to the Red Sea of Israel where he was taken aback by the fragrant nature of henna flowers growing alongside at night. He then attempted to obtain essential oil from the flowers to take back to the Campo lambs. Using the cutting edge essential oil extracting methods (EtoH and Co_2) he found that he was hardly able to obtain one drop of pure essential oil. Due to the extreme expense he saw would be needed to further collect essential oil from the henna, he quickly abandoned the project. Henna essential oil can not be distilled like other essential oils thus driving its cost through the roof. While the attar method is still the best for making henna perfumes / oils, the use of sandalwood as a base adulterates the essential oil and may lead to incorrect pharmacological findings. Many agree that henna's scent is best when mixed with the night air or burned as incense instead of bottled up. In Arabian areas such as Yemen, henna was the basis for an old and still popular incense which was said to have a blessing (baraka) effect when burned. It combined rose water (*Rosa spp.*), tea, sugar, henna oil, sandalwood oil (*Santalum album*), musk oil, amber oil and aloe (*aloe vera*). To recreate this incense simply combine the sugar and rose water until the sugar dissolves (make the ratio 1 cup sugar to 1 & 1/2 cup rose water). Once this is done, turn the heat off and immediately add the rest of the fragrant ingredients including henna flowers / henna attar oil and the gel of the aloe vera. This is then poured into a shallow dish and allowed to cool. Once cool, it should be crumbled and stored in an air tight container. To burn it, place it atop a charcoal incense burning disk. King(TM) is a popular maker of such charcoal disks and may be found in most stores selling incense or health-food / new-age products. Incidentally, "kyphi" was burned as incense in the evenings and said to lull one to sleep with its intoxicating scent. Every night it was burned in Egypt for Ra to create rest over the entire land. Henna flowers were also used in fresh form by women and dried form as well because they hold their scent extremely well, like rose petals and lavender for instance. Basically, little, if almost nothing has been written in regards to Aromatherapy and henna essential oil. There are but only bits and pieces, usually in regards to henna's use as a perfume or on a pure Olfactory level. Below is a compilation of such information and my own personal observations concerning henna essential for those wishing to use it in a therapeutic and perfume nature.

Lawsonia inermis:

Henna Essential Oil:
Synonyms: L. alba, L. spinosa, L. mimata, L. ruba Mehndi, Gulhinna, Mehndi Attar, Hinna Attar, Hinna Perfume

Herbal / Folk Traditions:
Henna has a well established tradition in folk medicine and its scent is frequently described as being both sedative and an aphrodisiac. People would inhale the scent from fresh flowers or the oil to relieve headaches, calm anger / frustration and to induce sleep.

Aromatherapy / Home Use:
May be used on the hair to treat balding and thinning hair. Inhaling or applying it to the forehead to relieve headaches. A massage oil may be created and applied to joints to relieve pain. A few drops on ones pillow may induce sleep.

Extraction:
1. Essential oil by steam distillation from the flowers. 2. An attar is created by infusing henna flowers into low grade sandalwood oil. 3. Henna flowers are soaked in non-scented oil for up to 20 days to extract properties.

Characteristics:
1. The oil is a deep red, to yellow, to orange, to brown in color, viscus in nature and has a very sweet scent which some note is almost fruity; it is likened to borania and jasmine with tealike, leafy undertones. It would be considered a top note and a floral. It blends well with most florals and woods such as jasmine, rose, ylang ylang, sandalwood, cederwood and patchouli.

Actions:
Sedative, anti-fungal, anti-bacterial, anti-inflammatory, allergenic (to some), uplifting, antiviral, antiseptic, deodorizing.

Safety data:
Safety data unavailable at present. Henna oil is not commercially produced and frequently adulterated with low grade sandalwood oil or simply added to other complex perfumes in crude oil form. The A-ionone in henna essential oil may cause allergies to flair up in some.

Chapter Four

Beauty is in the Kholed Eye of the Beholder

Henna's Cosmetic Life

> "They found that the crow was the false one and for that they painted him black as his lie. The partridge and dove were rewarded. Stinging the feet of the dove joyous red with henna and lining the eyes of the partridge with kohl. To this day the dove walks on red feet and the partridge's eyes are ringed with black."
> —Old Arabian Folktale

The cosmetic life of henna is too a rich one! Even though Mehndi staining or the application of designs using henna on the skin can be looked upon as cosmetic, we will be dealing with henna's use on the hair, face and nails predominately in this chapter. Mehndi is covered more in-depth in subsequent chapters and also my previous book "Mehndi: Rediscovering Henna Body Art", the latter containing recipes.

HENNAED HAIR

> "She's tinged her feet and hair with henna,
> She has tresses—her hyacinth auburn hair
> She praises herself every morning,
> Her golden hair reaches to her feet
> Her hair smells fragrant,
> Thin waist coquettish lover
> I shall wait years for you."
> —Dadaloglu 1850

By far, the most popular application for henna throughout history is as a hair-colorant. Both men and women employed it and few other natural substances could produce the vibrant effects on hair henna powder does. Some of the first people to use henna to the fullest was the ancient Egyptians. Due to their ingenious mummification process, archeologists and researchers are able to do DNA and other forms of testing to determine if henna, which many times is found present, was used on the individual in both life and death. Even before the advent of such scientific testing, archeologists have long noted the henna used to stain the hair, wigs, nails and so forth of the mummies. Even though myths swirl that an Egyptian queen was the first to discover henna as a hair colorant while sitting next to the Nile 3000 years ago, researchers believe henna use has been in existence for such purposes much longer in the area. This has come as a result of further investigation into the cosmetics created and used by ancient Egyptians. What is being learned is phenomenal. In the past, so much emphasis was placed on the actual mummy and any precious metal artifacts, that many times other items were glossed over and cast into backroom storage shelves. Just now the implements used in the mummification process are being tested on real corpses and the remnants of makeup in jars left in once sealed tombs, lab tested. Now, the ancient Egyptians are being credited with creating complex, chemistry like formulations for use as makeup. Recently cosmetic powders dating from at least 2000 BCE were found to be incorporating "wet" chemistry or otherwise the extraction of non-natural chemicals using various types of water solutions. These include the substances Phosgenite ($Pb_2Cl_2CO_3$) and Laurionite (PbOHCl) which do not naturally occur in nature. The powders, which were used for eye-shadow and eyeliner by the Egyptians, sat on a dusty shelf at the Louvre in France for over a hundred years before scientists began to test the contents. Like other aspects of Egyptian life, such cosmetics were actually thought to hold medicinal qualities including a "prescription eyeglass" effect to help one see better. Literally thousands of years later, the Greeks and Romans began to use the exact same substances to treat eye infections, trachoma and conjunctivitis. Of course these cosmetics were really caustic and toxic but it gives us a glimpse into the sophistication the Egyptians had when it came to cosmetic creation. It also shows how the Egyptians could have quite easily exploited every conceivable aspect of henna prior to other cultures. Henna, in order to dye the skin or hair, needs a bit of processing. It can't be simply picked off the bush, mixed with water and *presto*, you have an orange or red dye. In reality the henna leaf or berries / seeds needs to dry and be crushed into a powder to get maximum staining. Just like tea needs to be dried before it can be steeped in water for maximum taste. From extremely early on, both men and women in Egypt used henna for a number of purposes. One way for men was as a colorant for their beards. From the most early records, Egyptian man wore various types of natural beards, some of which were dyed with henna and even at times interwoven with golden threads. Later, false metal beards were worn by pharaohs as a sign of sovereignty. The coloring of hair, both natural and in wigs, has also been well documented. To date, some of the oldest mummies ever found (HK43 Naqada II period c. 3650-3300 BCE) in Hierakonpolis have been discovered to be wearing braided hair-weaves and natural hair which had the presence of henna dye. The henna was detected using lab technology. This latest archeological find would certainly dismiss the notion that the people of Crete introduced henna to Egypt in the 2nd millennium as proposed by Kenneth Dachman and Joen Kinnan. Also that henna first *began* to be used in the 16th century BCE as suggested by other authors. This find in Hierakonpolis really rewrote the history of henna as a whole, as no older direct artifacts have been found. Of course other colorants were also used including a black pigment prepared from manganeasem bicarbonate and quartz powder mixed with fat or oil to make a type of pomade. The henna on the other hand could have been used in either paste (mud like) form on the hair or via a wash (rinse) created from an infusion of henna leaves in water. Henna also has the ability to absorb oil from the hair, rendering oily hair to a more dryer, normal state. Lice, ringworm and other parasitic problems plagued the ancient Egyptians, causing some to rather go bald and wear wigs. So such an attribute of henna would have been very appreciated. Henna on the hair was likewise viewed in a medicinal way. According to the medical papyri (of Eber) from around 1500 / 1550 BCE, henna (henu) was used in a recipe to help re-grow hair and encourage it to become longer and stronger.

A 2nd century AD book by Parakarana II named "Navanitaka" appears to show henna was being used in complex hair-coloring formulations in India, including one that incorporated henna with numerous other ingredients; powdered iron, lotus stalk, mango, blue lotus, sesame oil, licorice, etc. In India, henna was most likely used for its red color on the beards of men. Black hair was more popular with women, thus the use of henna as an ingredient along with other colorants such as blue indigo to create a dark dye. Both Pliny and Deioscorides noted henna's hair coloring qualities where the latter said henna was a plant "whose leaves dye the hair of an orange color". At the time of both men (c. 60 AD) and even earlier in Theopharastus' time, people were just as fascinated with ancient Egypt as we are today. They were especially interested in the herbs and botanicals they used. With the creation of the Materia Medica's, the Egyptians use of henna spread especially among the Romans and Greeks. This was particularly true of the era after the death of Cleopatra, where the Romans adopted many of the ancient Egyptian customs. In Egypt during Roman rule, it became much like 19th century New York City America. A melting pot of cultures and religions in addition to a mecca for trade. Both Romans and Greeks used henna as a darkener for their hair, most likely after seeing the Egyptians repeated use of it. Jews, whom are also credited as being a clean people, would use the henna in their baths and for hair care in addition to other cultures living about the area. This is confirmed by many notations by the ancients that some of the best henna grew by the Dead Sea, Engaddi and the Ascalon in Judaea. Egypt was also noted for producing the finest henna, especially around the Nile and Nubia. This proliferation of henna use most likely shaped Mohammed's (632 AD) early life in Mecca, causing him to declare henna to be "best of herbs". Later, henna was devoted to his memory among the Muslim peoples. We know henna was already an important hair colorant prior to the birth of Muhammad and Islam in the area of Arabia and Mecca as a few poems have surfaced including the "hanged poems" of Ka'aba. One by Imru-Ui-Quais, found on an ancient temple wall in Mecca (622 BCE) has been translated as "The blood of many leaders heads is in him, thick as the juice of henna in combed white hair…". Muhammad is quoted as saying "Use henna, it makes your head lustrous…" and something to the effect that henna was the best of all hair dyes. From that period on, henna was frequently used as a colorant for Muslim men's beards and for women's hair in addition to other botanicals such as indigo, katam and Persian walnut. This is seen by a poem by Imr-Ul-Kais (540 AD) where in the Moallakat he writes, "As the henna juice lies dyed on a beard grown boar, like blood." In some areas, the use of henna in the hair was a marker of a woman being married or widowed. One example is the Ait Haddidou people who live in the mountains of Morocco. Since the peoples have been isolated for so long, their customs may be reminiscent of ancient days and traditions which have since been lost, in other areas. Akidoud, their name for the henna hair dye / mud, is applied to the hair of married or widowed women only, usually during festivities and cultural dances. Many cultures will reserve the use of henna for mature adults or after the age of 13. In most Muslim communities, women are not allowed to comb their hair or use henna during the morning period and especially if the deceased is their spouse. Even the infamous Karma-Sutra (320—540 AD) gives a hair recipe which is as follows:

> "The juice of the roots of the mandayantika plant (henna), the yellow amaranth, the anjanika plant, the clitoria ternateea and the shlakshnaparin plant, used as a lotion, will make hair grow…an ointment made by boiling the above roots in oil and rubbed in will make the hair black and will also gradually restore the hair that has fallen off.".

This is only one of two mentions of henna in the entire Karma-Sutra text, leaving one to surmise henna's main uses in early India was as hair dye. In India and many other areas, henna was more than simply a hair colorant, it was used to stimulate growth and prevent baldness in both men and women. Dr. Mike whom works for the Uzbek Embassy to Israel and has studied Central Asian history for 20 years recounted his experiences with henna:

> "Uzbeks, Tadjiks and, to less extent, Kazakhs and Turkmen used henna to make their hair light. Usually henna was

mixed with Basma (black dye) in different proportions to achieve more light to more dark color.

Henna was [also] used to make hair strong and to prevent a bald patch. Henna was used with fresh garlic. Crushed garlic was mixed with henna. This thin gruel should be applied to the roots of hair and the head should be covered. Rinse hair after an hour. I saw woman who used this method and I can say the method works, indeed."

Later during the popularity of the "Hamam" bath houses in Turkey, special rooms with bowls of henna were created to allow women to wash their hair. At times the henna was provided for them and at other times women would bring their own. In addition to giving the red highlights and glossiness henna is known for, it was thought by the women it would make their hair grow quite long and strong. Having long hair that was kept "up" in elaborate buns and such was quite fashionable at the time. Numerous paintings from the time document this including "Mother and Daughter at the Hamam" by Rafael (Istanbul Turkey) dated 1754. Of course these bath houses were actually created years before by the Roman's and later maintained by the Byzantines; the henna seems to have been most popular during periods of Muslim occupation. The baths were more than a place to bathe, they were a meeting place for all economic levels of women and men (but at separate times of course), haven for matchmakers, even a spa of sorts. Samuel G. Wilsom (1895) notes on his trip to Persia, where bath houses are also found:

"Another place of social resort and gossip is the bath house. Custom and religion require frequent cleaning. For the men, with their dyeing of the hair and nails with henna, scraping the flesh with tufa, etc., the bath is a frequent necessity and no less so for the women, whose hair-dressing, dyeing of eyelashes, etc., require so much time and attention."

Women would bring anywhere from 15 to 20 separate items with them to the bath. This includes towels (pestemal) to dry oneself with and wrap hennaed hair, a metal bowl for pouring water over oneself (tas), soap, a small carpet to sit on while undressing, mirror of polished metal, essential oil (attar) of rose (*Rosa spp.*) and of course henna which was kept dry in a bowl and then wet with water once at the baths and applied to the hair. Some also note that in an attempt to keep cool inside the hot bath houses, balls of henna would be held in the hands and rolled under the feet. By the era of the Hamam baths, many cultures of women and men, including those in India and the Mediterranean, where using henna as a hair-colorant. This is documented by John Baptist Porta (Giambattista della Porta, contemporary of Galileo) in his book "Natural Magick" (1518) where he references a recipe using henna to dye the hair red:

> 'I shall teach you [to dye your hair red]. There is a powder brought to us from Africa, they commonly call it Alchena [henna]. If we boil it in a Lye until it becomes colored and anoint our hair with it, it will dye them red for many days, that is indelible. But while you handle it, take heed you wet your nails therewith, for they will be so dyed, you cannot easily make them clean. So also we dye the tails and manes of white houses red."

Porta makes a very good point not to allow your nails to be "stained" with the henna when applying it to ones hair. At that time and also much further back in ancient Egypt the luxury of wearing gloves to protect your hands while coloring ones hair was not available. Therefore anyone coloring their hair with henna would have gotten the added bonus of colored fingernails and even perhaps stained hands too. In any event, Porta's book shows that henna was being used in 16th century Italy. It also shows that henna needed to be imported from Africa / Egypt, leaving one to wonder if Dioscorides was correct in saying henna grew abundantly on Sicily. He likely was mixed up (along with Pliny) in thinking that "ligustrum" and henna were one in the same, which of course they are not. Even though "Bloody Mary" (Mary the I of England) made red hair all the rage, the effect was achieved using various bleaching agents including Lye, rhubarb, honey and a lot of sunlight. Henna, was according to Culpeper, mostly used as an ornamental bush and for burn / wound treatment. The medicinal uses gleaned directly from the writings of Galen. Plus the Europeans already had their beloved Alkanet for all their red cosmetic needs which included hair-coloring pomades and powders. It doesn't appear until well into the 19th century that European women finally discovered the red hair coloring qualities of henna. Chemists had long known about its effects and even tested it in their laboratories. Much of the cosmetic chemistry done on henna was at the turn of the century and prior, with a few tid-bits of new information such as molecule structure added when a scientist thought about it. This is largely due in part to the wave of "chemical" based experimentation that started after the 1890's. Using natural substances such as herbs, henna, became old, outdated and even "taboo". Chemicals and "electric" devices became all the rage until now where a resurgence of natural based products are hitting the market. Few colorants were able to give the vibrancy henna did though. An old Chemists recipe book published in England (late 1800's) sheds insight into the use of henna at the time by women:

> "It [henna] imparts a reddish-brown colour, like auburn. It is now a favorite hair-dye and the following hints are available. It should be noted that if care is not taken all

kinds of colours may result, which will lead the chemist into difficulties."

Obviously, henna was unpredictable to a certain extent on the hair. While it can color, in shades of red the best, it can't lighten or bleach the hair. The red color can only blend with the existing hair color. Most people have variants in hair color, especially when white or gray is present. There henna dyed in these areas would appear much more vibrant and stand out from the rest of the resulting hair color. The same book offers tips on how to avoid such happenings:

> "Using henna, the powder is generally applied to the hair in the form of a thin paste as hot as can conveniently be borne. About 7 ounces of the powder is required. The paste is prepared by mixing the henna with sufficient, nearly boiling water on a water bath, on which it is allowed to stand for about ten minutes so that the coloring matter may be extracted before being applied to the hair. The mixture must not be allowed to boil. The paste is applied and the head covered in a paper turban. The length of time the pack is allowed to remain thus is important and depends on the nature of the hair, its original colour and the colour desired. Before and after applying the paste the hair is shampooed."

This method of application was used well into the 1960's. Powdered henna would come in small slat shaker type containers, many times pre-measured for one application or in a talcum powder canister. Many will remember actress Lucille Ball (I Love Lucy) and her large bottles of "Henna Rinse" to give her that vibrant red hair color. Ms. Ball was quoted as jokingly saying one…"I don't have an epitaph for myself…You know, it's nice to have entertained five generations…I never expected to be around this long and the length of time I've been around never occurred to me until one day recently, I found out I was out living my supply of henna.". Today, in the US, henna is only indicated by the FDA for use on the hair. The Health & Safety sheets put out by the US government warns though that henna "makes the hair brittle" so using it no more than once a month is advisable. The brittleness most likely is a result of the henna stripping ones hair of all natural oils. People with very oily hair *may* be able to use henna more often.

RED NAILS

> "...the long fingers were perfect and the almond-shaped nails had been stained with henna, as was the embalmers' fashion."
>
> —H. Rider Haggard, 1912 Smith & the Pharaohs

There is a reason why red nail polish has always sold so well. Henna and other substances have long been used to give the finger and toe nails a lovely orange to red color. The key to obtaining the color one would like depends on how it is applied. One can not necessarily use the "Mehndi" application for deep red nails. Much debate has come over how the Egyptians used henna in regards to their nails. Many texts suggest that not using henna on the fingernails was uncouth and improper. Others indicate henna was only applied during the mummification process. The latter most likely stems from the lack of pictorial "evidence" that Egyptian women used henna on their nails. The same could be said for the green ointment made from malachite used under the eyes. Only but one or two reliefs show it on an Egyptian yet bottles and jars with applicators have been found in some of the oldest tombs discovered to date. Further evidence that one can not necessarily place a lot of weight on the Egyptian art which is filled with riddles and doesn't always endeavor to depict reality. According to manicure historians, both Queen Nefretete and Cleopatra were depicted (while living) with red and rust colored fingernails. Texts also document that during at least Nefretete's era, only royal women could wear such bright colors on the nails. At the same time, a number of other nail colorants were also around including berries, Alkanet and also certain animal fats which the Greeks were known to use. So using art to prove the use of henna on the nails may create a somewhat distorted picture. Many archeologists have come to the conclusion the Egyptians simply dipped their fingertips in henna and that was the extent of the nail adornment. This is many times seen on reliefs and tomb paintings in the form of orange tipped fingers. One of the earliest found depiction's that may indicate such comes from the tomb of Ramose of Thebes (c. 1370 BCE). The mural shows wailing women attending his funeral with orange colored hands. Other depiction's may include the "Nebaumens Banquet" at the tomb of Neb-Amun (c. 1400 BCE), art from the tomb of Nakht (c. 1370 BCE) and the depiction of Akhrnaten's daughters (c. 1375 BCE). Unfortunately the darkened hands / feet can be the result of the red granite wall peeking through the paint, so the pictures can not be highly relied on. Some also try to point out that a lack of numerous depiction's of women wearing colorant on the hands equals a lack of popularity. To date only one depiction has been found of a woman apparently wearing rouge and only two of people wearing green eye makeup. Yet the cosmetic jars for such have been found in almost every tomb, suggesting mass use of such substances in Egypt. Likewise finding only a few depiction's of henna on the hands of ancient Egyptian women would not be a good gauge of actual use. Of course, if women, even in the form of servants, colored hair using henna, their nails &/or whole hand would have been stained. Especially the nail because it only takes a few minutes for the henna to penetrate and stain. Any cultural group which colored their hair as much as the Egyptians did, with henna, would have quickly learned of this added bonus. According to Jennifer Wegner of the University of Pennsylvania Museum of Archaeology and Anthropology, the Egyptians did stain their nails, among other portions of their body while living with henna. She added it just wasn't intricate as Indian Mehndi is today. One example is said to be a Scribe named Ani (1400 BCE) which was mummified and was said to have had henna stained nails / fingers. Ramses II (1320 BCE) is also frequently noted as having hennaed hair, nails, palms and feet in addition to his wrapping. Another way the Egyptians could have colored their nails with henna, is via the oil / perfume extracted from its flowers. As you will remember, henna oil is not only thick, it's deep red in color. In the late 17th century, rubbing red oil into the nails was a popular way for women to make them attractive. As seen by China's name for the henna flower "fingernail-flower" a colored oil could be rendered from henna as well. Victorian women in Egypt actually used henna flowers for nail dye. *L. inermis* and *L. spinosa* are particularly mentioned as having flowers that can be used to color the nails. It has been noted that mummies have been found with hennaed nails dating from as early as the 11th dynasty and well into the Ptolemaic and Roman periods.

During the embalming process, some archeologists (such as Haggard) noted that fake nails which were dyed with henna were tied onto the fingers of the mummy with thin strings. After tripping over the foot of a mummy laying on the floor, the famous artisan hired by Napoleon Bonaparte for his conquest of Egypt, Baron Dominique Vivant Denon (18th century), noted the following:

> "Undoubtedly, this foot is of such a delicate, elegant and able bodied girl in the prime of her life. Her soft feet have neither been scratched or painful walking nor nicked by rough shoes. Furthermore, she liked to color her nails with

henna in the way the Egyptian women still do nowadays, as they color their nails, hands and feet with henna."

He was so moved by what he perceived as a hennaed foot, that he stopped and painted it. He later included the depiction, with the above remarks in a book written in his native French when he returned from Egypt. His book is seen by many as one of the first sources of correct information to reach both France and England concerning ancient Egypt.

Once again, the use of henna on the nails seemed to be reserved for North Africa and Middle Eastern areas. This could have been due to the fact fungal infections were rampant. The Egyptians main nemeses was malaria, parasites and funguses which feed on human flesh and tissue. Sine the henna has anti-fungal properties and has been long used on the nails for such purposes by Arabians, it makes sense the ancient Egyptians noticed the effect as well. According to Unani healers, using henna makes the nails lustrous, grow stronger and prevents fungal infections. In the 16th century, a German Dr. noted henna's use for such purposes among the (according to him "white") Moors of Spain. Dr. Lang wrote (1526) about his visit to a well-to-do family:

> "In addition, with henna the young Moors of Castile obtain that their nails have a shining color of orange which (they imagine) gives a special attractiveness to them. In my opinion she [Moorish woman] makes herself resemble our miserable tanning of Nuremberg Germany."

Even though the Chinese were known for their royalties long, painted nails, they were lacquered instead of dyed. The Chinese were actually the first to create "nail shellacs" and they were quite found of the colors red, gold, black and silver. An early recipe for such lacquers included various plant pigments mixed with Arabic gum, egg whites, beeswax and so forth. Henna, known as fingernail flower in China could have been utilized for the red colored lacquers or for a nice staining effect. What is constantly shown though is that having long, painted or red finger nails was a sign of wealth. Most commoner people worked quite hard everyday just to survive and would not have the luxury of nice, "colored" finger nails. The exception could be on special occasions &/or gatherings. Still, if a woman did not grow henna herself and save some, obtaining it on a regular basis most likely would have been non-realistic. Buying food and clothing would have been more important, likewise using the henna medicinally. Various classes of Greek women are also thought to have stained their nails with henna which was kept in a toiletries box (Pyx). During the Middle Ages in Europe &/or Christianized areas, red became a sign of witchcraft and the use of such colors on the nails was tossed into the world of taboos. Of course they were most likely not using henna anyway but instead Alkanet. It still shows how the use of red hair colorants too would have been looked down upon. In Middle Eastern areas on the other hand using henna to stain the nails was alive and well. This is shown though the words of Muslims. One occasion Rasulullah received a letter / message on paper from a person sticking their hand through a screen. He remarked that he couldn't tell if the hand was a man's or a woman's, for which his guest noted it was a woman's hand. This led him to say, "If you were a woman, you would make a difference to your nails.", which was taken that she should use henna on them. This shows the mentality that henna on the nails and nails only was a popular female adornment. To add feminism and distinction. Just as with the use of henna on the hair predates the invention of Islam, so was henna used on the hands and nails. Some claim Fatima (daughter of Muhammad) was the first to discover henna as a nail colorant for women but this isn't supported and even countered by Islamic literature. In the tale of Mohammed's birth and his midwife (and life-long friend), a woman named Barakah, was separated from her new husband and sold into slavery, for which it was said the henna was still apparent on her hands. This was to show she had been married for but a few days and also brings out the fact henna was already in use during weddings according to the raconteurs.

Men would wear henna on the nails as well but mostly during battle and on special holidays such as Noruz. Samuel G. Wilson (1895) notes that every man in Persia a day before Noruz (New Years) bears in mind…"I must…dye my hands, nails and beard with henna…". Likewise in India, it was also popular to decorate only the nails with henna. This is gained from a passage from Dallana (c. 1100 AD) whom writes that mendi (henna) "…which the color of which woman paint their finger-nails.". Since henna was expensive to those who could not grow or harvest it themselves, many woman most likely applied the henna to the nails because they couldn't afford enough henna to do Mehndi on the hands and feet. It takes very little henna to stain the nails nicely. The nails being stained was also important for after a woman had a child in India with both the fingers and toe nails being decorated to purify. In India it has also been noted that woman substituted Chinese Balsam (*Impatiens balsamina*), one of the few plants to also contain the chemical Lawsone, for henna to stain the finger and toe nails. This would add extra explanation for why henna never caught on in China, as they could simply use their abundant native plant instead of harder to find and much more expensive to buy henna. Of course there were times when henna on the nails was not appropriate. Such as at funeral settings, after a loved one died (especially with Muslim peoples) and also on certain holidays and observances. This was noted by Sir Richard Burton in 1859 while on a trip to Mecca. Sir Burton frequently would dress himself in ethnic clothing in order to disguise himself which allowed him to write about his adventures freely. While in Mecca on a pilgrimage, he was told that he needed to change into pilgrim type clothing (Ihram). He recounted that he was told "…not to oil the body and not to cut our nails or hair, nor improve the tints of the latter with the coppery hue of henna. Transgression [sin] of these and other ceremonial exactments is expiated either by animal sacrifice or gifts of fruit or cereals to the poor.". What this also shows us is that henna is seen as a luxury and most likely has been for a very long time. A sinful act even for those trying to be pure during pilgrimage settings or pious in general. During the Spanish Inquisition (1478—1834) however the seemingly innocent use of henna on the nails turned into a criminal offense. Independent of the various papal Inquisitions which started hundreds of years before, the Spanish Inquisition was started by Spanish monarchs to punish the converting Muslims and Jews, in addition to driving out the ones already living there. One of the laws created to really hurt these two groups good was a ban on all use and sale of henna (c. 1520 AD). Prior to this however, in 1518, Emperor Carlos V made it a law that Castilians *only* could not use henna on the nails, as it stated Catholic Christians "can not put henna, neither the dust (powder) of the plant, on their feet nor their hands." Especially in a wedding setting as it appears that non-Moorish women adopted henna as part of their wedding ceremonies. These laws were obviously racially motivated. The women, many of which were not Jewish or Muslim complained about the law and countered it's use had nothing to with religion, that it was really only cosmetic. Of course in the case of the Muslims this wasn't actually truthful. It has been noted that ancient Islamic peoples used a mixture of water and henna in purification rites. The law was supposedly overturned but only a matter of a few years later it was enacted again. Women banned together to protest that using henna on the nails was not religious. A number of women were even put on trial for using it and used the defense henna use had nothing to do with religion. They obviously didn't understand the ban went much deeper than religious intolerance or separation. Many of the henna mills were owned by free people. Mills that were out numbering the royal mills. Mills that, in the eyes of the monarchs, were giving way too much economic power to the Moors. The laws seemed to be somewhat lenient at that time but at around 1566 the law was made permanent and strictly enforced, as seen by a number of women being put on trial for unauthorized use. Protesting that Christen women too used henna was futile as by that time everyone in Spain, including those not of the Jewish / Muslim religions were scared for their life as well. Being pulled in, tried and sentenced to life in jail or death was a real possibility. Women would rather give up henna then be tortured or killed. Being that this is so, we also see why perhaps Bloody Mary (who married into the Spanish monarch) didn't use henna to color her famous red hair. Although she died in 1558 she too would have been around during the ban and could have frowned upon its use in her England as well. Alkanet most likely was used in its place, it too can be used as a bright red nail colorant. The Spanish Inquisition lasted all the way until the 1830's. The ban on henna however does not appear like it lasted the whole Inquisition. It seems to have fallen through the cracks, as many ridiculous laws do. This can be seen in a passage of "Don Juan" by Byron (1819) which reads "Her nails were touched with henna" and goes on to say how it should be applied. It makes little sense for him to include such passages if women were not reverting back to their *old* ways.

OTHER COSMETIC USES

"...with kohl-darkened eyes and hands stained with henna dyes..."

—Sir Richard Francis Burton's translation of
Arabian Nights by Scheherzade

Depending on local, woman and men have found many interesting uses for henna cosmetically. Some we would never think to use today. One cosmetic and medicine in one was a distillation of the henna flowers which was used by ancient Egyptians to keep the skin supple. Salves and ointments were also created for similar use and as a natural sunblock. Due to the hot, extremely dye winds constantly blowing through Egypt, the use of cosmetics became more than simply beauty for the Egyptians. Without their fragrant henna products, which have built in sunscreen, skin cancer would have been a very great problem in addition to other afflictions from the hot sun. The Egyptians are said to have decorated their hands, feet and chests (bosom on women) with henna dye. Most likely for a combination of reasons including beauty and for added cooling purposes / sun protection. Having henna stained into the skin would most likely filter out more sun than simple infusion of flowers. Many of the clothing worn was quite shear and wouldn't do much to protect one from the rays of the sun. Depiction's of the body peeking through clothing, as if it were wet with oil, has led many archeologists to assume the Egyptians would oil their entire body daily in attempt to deal with the extreme conditions.

The Chinese on the other hand used henna cosmetically for *very* different reasons. In addition to staining the skin and oxidizing to various shades of orange—red—brown and even black, henna has been found to contain melanocyte stimulating hormones. Otherwise henna stimulates melanin production in the skin. This explains why Chinese women would use henna to cause their skin to become further golden in color. In order to achieve this, only the galls of the henna (Imperial galls) leaves were used. Like most galls, the henna galls contain more Lawsone and higher amounts of tannins, which the Chinese women thought was the be-all-end-all beauty treatment. Decoctions of the galls were created and the liquid applied to the skin. Instead of turning dark orange—red—brown like Mehndi patterns (which are created with dried leaves), the liquid would stimulate melanin production, resulting in a more golden complexion. In addition to this, it would act as a sunscreen as the ancient Egyptians used it. Throughout China's history, fair complexions have been a sign of feminism and wealth. Hence the use of henna by such wealthy women, commoners and those not living in Imperial palaces most likely had little if any access to henna. Today, cosmetic companies have rediscovered the natural melanin stimulating qualities of henna and extract the Lawsone for addition in sunless tanning creams and lotions.

Another use for henna which doesn't sound all to appealing is as a lip tint. Especially during the turn of the century, Cleopatra among other royal women were said to have tinted their lips with henna. In a line from the classic "Intolerance" a girl from Babylon is described as having "lips brilliant with the juice of henna...". Of course the Babylonians probably didn't ever use henna but such notations by novelists and writers at that time was frequent and probably spurred on by discoveries in Egypt. Many scoff at this though. Well, if people were drinking henna for medicinal reasons and a whole host of other horrible tasting items, why not use henna on the lips. The stain wouldn't be achieved using a paste like Mehndi does. That just isn't feasible. Perhaps a thin decoction / infusion of henna leaves were used. This was, unlike with powder, the leaves could be strained off and only the liquid (or juice) would remain to be applied. It may have tasted bad but the stain could have been instant. Not only that but like the Chinese galls being used to make Chinese women more golden in color, the henna would stimulate the lips to be darker and more naturally colored as well. Henna didn't have to be the only ingredient either, most likely other items such as berries, etc., were also combined. The Karma-Sutra by Vatsayayana even recommends henna as a red lip color:

"The color of the lips can be regained by means of the madayantika (henna) and other plants mentioned."

This is actually only one of two places henna is mentioned in the entire Karma-Sutra. Feeling the urge to emulate John Baptist Porta, I decided I would attempt to see what would happen if I listened to the Karma-Sutra and Cleopatra historians. Most responsible henna selling companies stress not to use henna on the delicate lip area, which I fully agree with but I wanted to separate myth from fact. Can henna be used to redden the lips? Or were all of the mentionings pure metaphor? I would soon find out. Taking some dried henna leaves (from my own shrubs), I used boiling water and lemon juice to make an infusion (tea). A bit of berry juice was also added. I then allowed it to cool slightly and strained off the leaves. The color of the liquid was a strange taupe. Using a cotton swab I applied the liquid to my lips, trying not to let it enter my mouth. Right away I noticed a bit of darkening, mainly on my top lip. The taste was pretty sour and unpleasant. After letting it soak into my lips, I left in on overnight. The next day I rushed to the mirror to see what, if anything, had happened. Well, not much. My lips did look posy but they were not bright red. I have a feeling I diluted the henna with too much catalyst. A stronger infusion maybe would have produced better effects. I'm sure my lips were well UV protected, even if they were not carmine red. It then dawned on my that women could have used the oil / perfume of henna flowers to get brighter effects. If you will remember, the oil is thick and deep red in color. Oil was a popular way of reddening the lips and since it isn't noticed how the henna was prepared for use on the lips, in the Karma-Sutra and other texts, it could have been a real possibility. Making the lips fragrant was also extremely popular with the Egyptians and this certainly would have been the case with henna oil. Another way to make a vibrant red lip tent is with Alkanet. When mixed with fat or oil, it produces a very vibrant colorant, which the ancient Egyptians were noted as using it for such.

Since much of ancient history was recorded by men and most makeup was used by women, a gender gap is created and certain things can become incorrect. One example of this is in "A Confederate Soldier in Egypt" where the author William Loring writes about how women use henna to stain the area around their eyes an orange color. This is the first I have ever heard of such a thing. I know that saffron and turmeric have been used for such purposes but henna in Egypt? Well, a few sentences later he exclaims that women of Victorian era Egypt obtained their henna designs via the use of henna flowers from the henna-tree. At first I thought this sounded incorrect, until further research provided that the henna flowers of certain species actually do produce a dye for cosmetic and textile use. This included the white flowers of *L. inermis* and *L. spinosa*. Back to the subject of henna around the eyes, the 1880 book "Lands of the Bible" also speaks of women using a red / orange makeup around the eyes but does not make a connection to henna. So perhaps that actually was a mistake of the author. Buck Whaley wrote in 1797 that Turkish women used henna on their eyebrows, so perhaps this is was Loring was going for. Returning to the subject of lip color, henna never would have caught on for such purposes in Europe as in 1770 a law was passed in England preventing women from wearing lip colorant. The law seemed to be targeted at "women of the night" and gold-diggers as it stipulated that any women found to have seduced a man using makeup and painted lips would be tried for the all too popular witchcraft. Laws like that would allow bobbies to add extra offenses to women who were picked up on another charge. A little bit later in time, an Essex woman in England was said by one source to have been tried for using "red ocher" which the historian explained was henna. I highly doubt this was so, henna and red ocher are far from the same. The woman most likely was trying to paint her lips red with the ocher on a regular basis, perhaps she was a performer or such and was picked up. Upon search of her house, they later used it against her as an extra offense. Unlike henna, the red ocher is not as permanent and could be rendered in fat to "paint the lips". Who knows, it could have been powdered Alkanet even.

Henna has been used as a toiletry for thousands of years. Included in baths by Jews to prevent and treat leprosy, rubbed on the skin as a fragrant oil by the Egyptians, held in the hands of a woman in a Turkish bath house. In addition to perfumes, the ancients also used plants that had absorbing and deodorant properties. Henna has such qualities and has been rubbed on the feet to cool and prevent / treat foot fungus (Athletes Feet), which was then known as "burning feet". Leaves still left on the branches were held under the arms to keep one cool and to prevent odor. This was noted by Shabeni's (18th century) translator in the early 19th century:

> "A decoction of the herb henna produces a deep orange dye. It is used generally by the females on their hands and feet; it allays the violence of perspiration in the part to which it is applied and imparts a coolness."

The leaves have also been used for such purposes loose in clothing as well to keep them fresh. Another way henna would keep one fresh is in the bath. Siddha physicians note that the leaves and flowers of the henna plant may be used as a detergent for the body and hair. In 19th century Persia, it was noted both men and women would massage henna all over their body and then rub it off with a tufa or cloth. This would have accomplished a natural exfoliating effect as well as left sunscreen properties on the skin for protection. In some areas, women would also affix fresh henna flowers to their hair or in their bosom to be naturally perfumed and deodorized all day. William Loring (19th century) was one of the men to recount such uses:

> "They [Egyptian women] love perfumes; it's a matter of deep delight in their everyday life to inhale the odor of attar and sweet smelling flowers. But of all of these the most agreeable to sight and smell is the universal henna, which diffuses its odors and embellishes every garden, however small. Like the lotus in the case of ancient Egypt, the flower of the henna is valued by the modern Egypt. The ladies carry it in their hands, perfume their bosoms with it and offer the beautiful flower to their neighbors. They are never failing companions in their homes. The significance of the flower is that of the emblem of fertility.".

Another important time in a persons life to use cosmetics was in death. The Jews would many times rub the body of a deceased with fine and expensive oil prior to wrapping them in gauze and subsequent burial. The Egyptians too did so but in of course a much more complicated manner. Archeologists are just now starting to really investigate how the embalming rituals seem to have been done. The discovery of the use of henna as part of the embalming rituals seems to have appeared in the 18th century by explorers, where it was noted in one journal, mummies were found with red dyed nails. According to them, henna was applied to both hands and feet during the beatification preparation of the mummy which took place in a special room called a "per nefer" which has been translated to mean "House of Beauty". The jury however is still out on the question if henna present was applied after death or if it is indicative of repeated use during life. This is due to the fact direct reference to henna being used during mummification by Herodotus is not made; unlike the case of frankincense and myrrh for example.

One of the stranger uses presented would have to be that of the Vietnamese who were said to create a concoction of henna, which they call Danh-to and other black ingredients to stain their teeth. Apparently, like certain parts of India, blackened teeth was quite in vogue. Hennaed faces however, on purpose, never seemed to be very popular. Statues from Athens and Crete were supposedly found with henna designs on their faces but unless henna was actually present on the artifacts, I find this highly unlikely. Henna does a *very* poor job of staining the face. In experiments I did, a very faint orange only appeared which dissipated in a matter of a couple of days. It was barely even visible. The reason for this is the face is very oily and has many pores. Both oil and pores are the arch nemesis of henna dye's ability to adhere to the protein of the skin. People of India, especially laborers have been known to splatter henna on their faces as a sunblock but they leave the paste on, like a mud mask. Had henna really been used for a nice red color on the face, women of India would have tossed their kumkum long ago for a longer lasting hennaed bindi. All references to ancient statue artifacts having hennaed faces (or anything else for that matter) is pure speculation. In reality it was likely something painted on or representing permanent tattooing or even something purely iconographic.

Chapter Five

✤

The Lepers Only Hope
The Medicinal Use of Henna Throughout the Ages

"A pitcher of mignonette
In a tenement's highest casement-,
Queer sort of flower-pot-yet
That pitcher of mignonette
Is a garden in heaven set,
To the little sick child in the basement-
The pitcher of mignonette,
In a tenement's highest casement-"
　　—Henry Cuyler Bunner 1895 A Picher of Mignonette

HENNA'S MOST COMMON MEDICINAL USES

NOTE: *The purpose of this chapter is to provide fascinating and historical information to the general public concerning the medicinal use of henna. In presenting this information, I neither suggest not recommend its use. I especially do not recommend it being taken internally or ingested in any way whatsoever. In addition, the recreation of the ancient formulas found here is not recommended to be done by laypersons. I, nor the publisher except any responsibility for the actions of the reader(s) in regards to this chapter or the entire book.*

Poison (documented in Mexico)
Soreness and Pains
Headaches
Boils (on the skin)
Bruises
Burns
Burning Feet (now known as Athletes foot fungus)
Candida
Deodorant
Condyloma
Dermatosis
Fevers
Herpes
Hoarseness
Inflammations
Hysteria
Leprosy
Jaundice
Eczema
Lecorrhea
Scabies
Onychosis
Myalgia (documented in India)
Parturition

Ophthalmia
Ulcers
Purgative
Sores
Cancerous Tumors
Venereal Diseases
Sore Throats
Dandruff
Whitlow

Note: Conditions are in no particular order and come from traditional documented use.

Astringent
Bactericide
Poultices
Sedative
Vermifuge (expels worms)
Fungicide
Plaster
Snuff
Diuretic
Infusion / Decoction

Another band of misinformation being spread (especially among Mehndi artisans) is that henna has absolutely no medicinal qualities. That it purely acts on a placebo plane or something to the nature of folk superstitions. This is because few have really investigated its medicinal roots. So much emphasis is placed on its cosmetic value today that its medicinal contributions to the world have been pushed to the way side, especially by Western Mehndi artisans trying to place an exaggerated measure of sacred exclusivity to its artistic side. In reality, henna has shown many interesting medicinal qualities, especially in the areas of cancer research, Candida, Fibro Myalgia and bacterial inhibition. In many areas of the world such as the Bulkins, it was used purely for its medicinal value before its later adoption into cosmetic use. Once again, one can think of the Egyptians and their views of perfumery. To them, it was medicine even though today we may see it as cosmetic or perfume use. People of India are frequently doing studies on the effects of henna on a number of conditions which will be detailed later on and finding surprising results. In reality henna is a major part of many herbal based medicines and saying it has little or no medicinal importance degrades these ancient healing techniques. Medicinal texts, many quite old, hold a wealth of information about henna including its myths and insight into other aspects of its traditional uses. Since these texts were written to represent their time or document the past, they are also some of the most credible historical sources pertaining to henna. It also shows how intercultural henna use really was; another reason some try to erase this areas of henna's history.

Most definitely, there is a large amount of documentation that the Egyptians had an advanced capability of doing distillation and early prescription type medicinal techniques. Henna most likely was first extensively used in essential oil form as other botanicals were such as myrrh and cedarwood. Due to the horrific heat and sun of arid Egypt, oiling ones body constantly was very important. To the higher classes that is. The servants and lower classes of laborers (slaves) many times did not have access to essential oils. The use of the body oils would keep their skin supple and with botanicals such as henna, actually protect them from the UV rays of the sun, acting as a sunscreen. Researchers have located and identified chemicals inherent to henna which actually allows this to happen, Fraxetin and Aeculetin are two of them. The Egyptians also frequently used tannic acid as a medicine and to help heal burns and other skin conditions; since henna too produces tannin it could have been used by the Egyptians for that application as well. Once again modern science many times will validate the uses of botanicals that have a long tract record. Unfortunately, as described in the chapter on the aromatic use of henna, the essential oils were seen less by the Greeks and Romans as medicine and more for pleasure. This most certainly had a negative effect on henna's medicinal history in those areas. The Egyptian medicine of ancient times seems ever more sophisticated as archeologists take a closer look at their artifacts and papyri. The Egyptians though most diseases were brought on by evil spirits, hence the use of incense, internal decay and worms. One of the most interesting findings still being explored is a pin that was inserted into the knee of a mummified man. All sorts of surgeries are being investigated including that of the brain even. The most interesting discovery is that of an actual woman doctor by the name of Lady Peseshet (Late Kingdom—2600 BCE). Peseshet, which means "overseer of the doctors" is believed to have been the first woman doctor ever. Another name for her has been "imyt-rhm(wt)-ka" meaning "the soul-priests" in relation to the women doctors reserved for mummifying female bodies. Hundreds of other female doctors who came after her have also been documented in Egypt. No such positions for women have been found in any other ancient civilization including Mesopotamia. What is also quite evident is the Egyptians had a vast knowledge of herbology. While some of their concoctions were quite disgusting sounding, many of their treatment notations are from trial and error as opposed to just myth induced theories. Important papyri texts which give a glimpse into their use of healing botanicals include the famous "Papyrus Ebers" (c. 1500 BCE but some believe it is copied from much older papyruses from around 3300 BCE). For which henna in addition to other well known botanicals are included. What should be pointed out is the information provided in these papyri indicates a historical perspective. Otherwise the person writing the papyrus was attempting to preserve the *long* historical use of the botanicals mentioned. Even long ago other cultures marveled over the Egyptians knowledge and practice of medicine, spurring Homer to write:

> "That fecund land brings forth abundant herbs,
> Some beneful and some curative when duly mixed.
> There, every man's a doctor; every man
> Knows better than all others how to treat
> All manner of disease…"

He was actually very correct because just like every pagan Egyptian household had a practicing priest (usually the head member of the family), it was very popular to study medicine. Everyone, especially kings and others with wealth studied medicine. Hence the Ebers papyrus which is considered a textbook of Egyptian medicine.

Throughout time, in these same areas, Leprosy was a horrible and wide spread problem. So much so, that areas far away from normal civilization was created to warehouse afflicted people. They could be likened to concentration camps. There were also very strict laws created to protect the healthy public from infected peoples, including burning down and destroying the persons house and belongings and carefully wrapping the person in gauze to try and keep infection minimal. At that time, little could be done for such people. Henna, made into a paste and applied to the sores to keep infection minimal was used, especially by Israelite peoples. While the henna could not cure Leprosy bacillus, it could help stop the bleeding and inhibit infection of the open sores. Found to have astringent and bactericidal properties, this would explain its use for this serious condition. At this time, henna was surely valued for its healing properties and was probably used in both essential oil and herbal form. Extracted from tub text, pharaoh speaks about trying to be healed of heat exhaustion by going to the land of henu (henna) which is said to have been biblical Goshen. The use of henna to cool ones body and cure headaches from the heat is one you will find mentioned in many areas of Egypt, the Middle East, Africa and India. Once again this can be validated by researchers finding henna has mild analgesic (as in Aspirin) properties and would account for such use. Many times the henna was formed into a thick, clay like, paste that was held in the hands as a ball or laid across the forehead. When henna powder is mixed with oil, it will not impart color to the skin. Branches of it were even held under the arms, all in an attempt to cope with the extreme heat conditions. Anyone whom has had a chance to experience real henna paste will note that as soon as its placed on the skin, it feels cool. One thing to remember is henna grows in plain areas where the sun is hottest and it has to protect itself from burning up or its leaves becoming dried out. Many peoples found these attributes could be transferred to them when the henna was powdered or an infusion created.

Most common medicinal uses of henna in ancient Egypt:
—For scorpion stings and snake bite
—Headaches, especially from heat
—In Khphi, which was used to induce sleep
—In oils and ointments to soothe and protect the skin
—Sunblock
—Astringent
—Hair growth / baldness tonic
—Close open wounds
—Heal burns

MEDICINE SEPARATES FROM MYSTICISM

In ancient times there was little separation or concept of health, religion, the cause of illness and so on. Most thinking, including by physicians of the time, was that past evil deeds or evil spirits caused one to become sick or that one broke a bone. Being wounded in battle or braking a bone was seen somewhat more as a real affliction than lets say getting a fever all of the sudden. That was more of the realm of evil spirits and other superstitions. Preventative medicine was very unheard of with the exception of Biblical Jews who created a number of smart laws including burying human waste far from camp, making people with Leprosy call out "unclean", banning close marriage and so forth. Many feel their form of prevention made an enormous contribution to medicine as a whole, as other cultures of their time and even afterwards didn't appear to do the same. This thinking slowly changed especially with the more advanced dynasties including 2500 (or some say 1000) BCE China and also Vedic medicine. While much superstition was still attached, scholars began to document which botanicals worked and which did not, also for specific conditions and how they could be used to balance the body. Many of the ancient texts are still used today and the basis of these important traditional medicines. It took a while for the Greeks, whom are said to have been importing henna to their country since 1400 BCE, to come up with the same philosophical basis for their medicine but when they did, it swept the continent and much of the world too. Hipprocrates (486 BCE), thought by many as the father of medicine, did much to start categorizing botanicals and foods that could balance the body and prevent disease. The model, which looks very much like the Chinese, Japanese, Ayurvedic, etc., includes cold, dry, damp and hot categories. When the body falls too much into one of these four areas, the thinking was one would become ill. This shows the roots of preventive medicine. Trying to do something medicinally *before* one becomes sick, not afterwards was a major step for medical thinking and practice. Later on Pedanius Dioscorides (60 AD) wrote one of the first herbals called "De Materia Medica". Dioscorides, whom was thought to be an army physician, created very detailed passages on the use of botanicals, food items and on specific conditions. He spoke of henna as a plant "....whose leaves dye the hair of an orange color.". He also said he could make a red dye from it like "the blood of a dragon". In Dioscorides' time and local, henna was called Cyprus (kypros / appelle cyprus) and he recounted the following..."the best quality grows at Canopus in the Delta of Egypt". He also noted "...henna is not found everywhere and it must be ground and prepared into ointment and thus carried to other countries and regions.". Both he and Pliny Elder (60 AD) created fragrant unguents (ointments) from the henna which was stored in alabaster jars and used for a number of conditions. These henna ointments were said to keep for 3 full years. Unguents were the mode of choice for administering medicine in that era. According to Dioscorides, henna is useful for treating female parts, perhaps meaning aiding a woman's menstrual problems. He also spoke of it being warming and calming and henna ointment and oil was added to medicines intended to calm yet prevent weariness. An example of such an unguent may come from an ancient Parthian recipe which combined henna with cinnamon, spikenard, saffron, cardamom, cassia, myrrh, calamus, marjoram, fine wine, honey and a few other exotic ingredients. This was then added to fats to create a thick ointment.

The Greek form of healing reached Rome in about 100 BCE. Very popular and lucrative to doctors, the original writings by Hippocrates was reformulated by Claudius Galenous or Galen (170 AD), inventor of cold cream and deity to modern pharmacists the world over, which had a huge effect on medicine. His books were copied, many times by monks and carried all over Rome, the Middle East, Europe, etc. In addition to his creation of the theory of humors controlling health, he also laid the foundation for "The Law of Signatures". Signatures were used to identify the healing properties of plants just by looking at them, where they grew, atmospheric conditions and so forth. Thus, Galen's use for henna quite differs from the writings of his forefathers, although he may have extracted something from the writings of Hippocrates. He also admitted to learning a great deal from Imhotep's temple in Memphis Egypt so his uses probably mimics that of the ancient Egyptians as well. One of the uses for henna according to Galen is to stop the dilation of the pupils of the eyes. A more familiar use he notes is for healing burns from fire. He also states it is used for ulcers (sores) of the mouth, just like the Chinese used it for. Practically all of Culpeper's notations on henna's use in his time are extracted from the writings of Galen. Galen's recipes for henna are quite simple, such as mixing it in animal fat and applying it. It was also used for people with too much heat in their body and as the familiar "cooling substance". Henna certainly would fall under Galen's "cold" category. By the time of Galen's death, medicine was said to have been as advanced as 17th century Europe and had an extreme Egyptian flavor. When the Roman empire fell in the 5th century, a great shifting of educational learning happened. It was no longer seen as the empire of enlightenment or the place to seek an education about medicine / herbology. Instead, the Near East, namely Persia was teaching the wonders of Galenical medicine. This had a profound effect on the Muslim peoples who enthusiastically adopted it and combined it with their previous Egyptian and Arabian folk medicine. Galenical medicine was now known as Unani Tib. What needs to be pointed out is both Persians and Moguls practiced this type of herbal medicine. This is what led to India's adoption of it into their medicine as the Persians invaded and the Moguls in the 14th century. This has resulted in the majority of henna use in India being extremely reminiscent of Unani Tib teachings. Even thought Ayurveda is the ancient and first true form of healing in India, it was replaced and later influenced by the Unani Tib and still what is primarily practiced in India. Ayurveda is making a resurgence there, just as it is being further discovered in the West. When cultures take over one another, many times the traditional form of medicine is destroyed or must go into hiding, thus merging with the newer forms of healing taking over.

HENNA PROVES AN IMPORTANT MEDICINE IN BATTLE AND ISLAM

> "Use henna, it makes your head lustrous, cleanses your heart, increases the sexual vigor and will be witness in your graves."
>
> —Muhammad

One interesting account shows how henna really was an important medicine in the 13th century. In 1254 AD the inhabitants of the castle of Mihrin (said to be not far from Girdkuh and Shahdiz in what is now Iran) was being destroyed by a massive outbreak of cholera. This was extremely devastating because this was a stronghold of soldiers trying to hold back the invaders. Hearing of this, Alauddin Muhammad decided to send reinforcements in addition to medicine for the ill. 110 men marched on foot to the castle each carrying three loads of salt and two loads of henna apiece. Although none of the herbals or medicinal literature of the time recommended henna for curing cholera, the soldiers of the castle made a special request for it. It turned out, when they were running out of clean water to drink, some began to drink henna water. The ones who did found themselves cured of cholera. It isn't known if they were making an infusion of the henna leaves or mixing powdered henna into the water but which ever it was, once the reinforcements came with the henna, the cholera outbreak ceased and they were able to hold back the invading army for another few years.

Of that era henna was most often used for the same things the Egyptians were known for; headaches, sunscreen, bleeding, skin problems and so forth. It was also used frequently as a form of protection for men in battle. More in a superstitious manor when applied to the nails and hands but still most likely used on battle wounds as well. Unlike other cultures armies, the Muslims frowned upon using branding as a form of closing or cleaning wounds. Henna on the other hand was quite all right and could form a protective cover over the wound in the form of a plaster and serve to expel fluid. Something that the Muslims were very adamant in doing when it came to wounds. Fluid (or puss) was a sign of continued infection and needed to be removed before healthy tissue could fill in its place according to ancient texts. Very smart thinking actually. The Islamic medical texts speaks much about the way henna was used back around 651 AD and onward in traditional medicine. Muhammad was quoted as saying henna was "best of herbs.". Both men and women were highly encouraged to use henna for a variety of applications, both medicinal and cosmetic wise. Men were especially encouraged to use henna as a colorant for their hair and beard and also as an aphrodisiac. One passage reads, "no injury or thorn piercing was treated on which henna was not applied."., showing that henna was used both on minor and serious wound conditions. In addition to this, just as with the Egyptians, henna was used for headaches, leg pains, scorpion stings and Leprosy. It was also used internally in the form of a decoction of the leaves which was prescribed for a length of time for skin conditions and hemorrhoids. A special concoction of henna, rose oil and bees wax, which must have smelled marvelous, was created to be used on boils, skin abscesses, burns and aching muscles. Dried henna leaves were even placed in clothing to repel bugs. Researchers have found a number of inherent chemicals of henna that act as natural insect repellents including Gallic-Acid and Mannitol. A thick paste was used on the soles of the feet during Small-pox outbreaks and said to prevent the eyes from being effected and the sores to heal faster. Henna was associated with the cold and dry humors. This was intern linked to the element of Earth (the other three elements being air, fire and water) according to Galenical humor charts. This would also explain its use for cleansing of the blood and liver as Melancholic temperamented people were though to lean too far towards being "cold and dry" and thus more prone to the aforementioned conditions. It was thought that being in perfect balance would prevent sickness but most people were thought to always lean in one direction and temperament type (cold and damp, damp and hot, hot and dry or dry and cold). Therefore henna was most likely reserved for people in the cold and dry realms of health. As noted before, the writings of Galen and his Galenical medicine had a profound effect on Middle and Near Eastern peoples. One of these people who was an important figure in medicine for almost 500 years was Avicenna Ibn Sina (1037 AD) of Persia who wrote the monumental book, "The Canon of Medicine". The bases for Unani Tib, Avicenna brought all of the major medicines of his time together in his book including Egyptian, Persian, Greek, Roman and so forth. He especially dwelled on the writings of good old Hippocrates and Galen. He also of course incorporated henna into his books content. Resembling the medical models of Hippocrates and Galen, except in a more complex form, the writings of Avicenna do much to enlighten us on the ways henna was seen in not only his time but a good 500 years later, all over the Old World. Firstly, like Galen, Avicenna placed henna in the category of being "cold / cool" or "sardi". This then related, according to his charts, to the element of "Earth". This then led back to Melancholic temperamented people as the main ones that would be in need of henna. The element of earth, which henna could be called an earth herb, was seen as being a heavy element like water. Fire and air were of course light. Such heavy earth herbs were attributed to strong, negative, passive and female people. Showing that henna was considered a good herb for women. Going on further through his charts, one sees that earth herbs and henna are indicated for *mature* people, not children, the season of fall and the humor of black bile (sauda). So another thing that may be picked out is henna would be indicated for adults, not children and perhaps used mainly in the fall season. Going on, earth herbs were said to have the following effect on human physiology:

Excretion: feces
Tendency: spreading
Sense: touch
Mentality: torpid
Mental state: obstinacy and fearfulness

Yet again we can pick out the "touch" and comforting in relation to henna and torpidity for people whom would need henna as an aphrodisiac possibly. We also once again have the adult and woman connection as well. This all has implication on how henna was used by women as not only a medicine but also in the manner we think to be cosmetically today. In order to be placed in the "cold" category, one had to inspect how henna was being used in the past as well as the present. This all comes from the Greek model which linked the body to the earth as a whole.

Common medicinal uses of henna by Muslims included:
—Sore muscles and joint pain (Arthritis)
—Wounds including insect bites, scrapes, ulcers and bruises.
—Male stimulant (aphrodisiac)
—Astringent and anti-hemorrhagic
—Fevers and headache
—Chicken and small-pox

THE LAW OF SIGNATURES BECOMES THE LAW OF THE LAND

Greatly aided by the Persians and Muslims adoption of Glaenical medicine, which was known as Unani Tib (Tibb), a whole new form of healing begun. Not only this but more of an appreciation and interest in botanicals. Since now healing plants were separated from food plants and poisonous plants, more emphasis was placed on finding these curative herbs. Identifying them, recording them and testing them became very important to physicians and scholars. Unfortunately this is also a time when some botanicals began to be used for conditions they really could not treat. Turned into "cure-alls" or had their real healing qualities inflated. This is what subsequently happened to henna. Instead of just being used for sunscreen, headaches, wounds, etc., it was now a cure for jaundice too. Also it was touted as a blood and liver purifier. All sorts of new applications for its use seemed to jump out. This is most likely due to the fact new healing actions were being attributed to henna as a result of the "Law of Signatures", also referred to as the Doctrine of Signatures. A very complex system, still taught and used, to a certain extent by master herbalists, the Law of Signatures is supposed to allow one to ascertain the healing qualities of a botanical purely by sight. Brewer noted:

> "According to the Doctrine of Signatures, nature labels every plant with a mark to show what it is good for."

Main areas to look at include a plants flowers, if any were present, the branch or steam formation, leaves, color of fruit, soil and growing conditions and the list goes on. In ancient times, most healing botanicals had to be wildcrafted or collected from nature and needed to be identified. Henna most definitely would have been associated with yellow, for a number of reasons and this would in turn explain its use for all of these new health conditions. The flowers of *L. inermis* and *L. alba* are anywhere from a white to yellow color. Yellow flowers according to the Law of Signatures means a plant can be used for purifying the liver, gallbladder and urinary track problems. It is also indicated for riding the body from toxins and infections. In addition to the henna plant having yellow blossoms, when the *fresh* leaf is torn or broken, a bright yellow liquid oozes out. Most likely this also tied the plant to the color yellow and the use of it with cases of jaundice. This could explain why some cultures, such as the Bedouin, would apply henna to newborn babies…which I might add is extremely dangerous. The henna has been found to do absolutely nothing to prevent of halt jaundice and can do much more to harm the baby than good due to its oxidizing and sensitizing attributes. On the other hand, *L. ruba* which can have all sorts of brightly colored flowers (red, purple, magenta, etc.), flowers could have been used more for its blood purifying qualities. Red has always been a sign of blood and a plants astringentness. The flowers pungent aroma also held clues to its medicinal value to early physicians which included being indicative of strong antiseptic and germicidal qualities. The leaves on the other hand, due to their somewhat soft nature, may have contributed to its use for skin and joint inflammation. Some of the Signatures of henna prove true while others did not and most likely, while still listed in books, were not used for conditions they were found not to treat. Except in the case of a last resort. The Law of Signatures is also a direct link to Medieval Alchemy as many of the Alchemists used books translated from Galen's works and while they might not have had access to all of the ingredients, there is a good chance that they did at least know of henna. Well into the 16th century, Galen's writings can be seen influencing the herbal and medicinal texts of the time. Going back to the 6th or 7th century AD in the Balkans there has been evidence found that Slavs brought there were introduced to henna either by

Arabian merchants traveling through the area or Christian missionaries. Most likely it was the Arabians but more interesting is how the Christians were also exposed to it and may have brought it to other areas of Europe as well. Many people were under the thinking henna wasn't used in Europe until the late 19th century but now written accounts of its use have been found in South Slavic Pharmacopoeia texts from the 13th century. It apparently was used mostly for hair care and not a dye for the skin. The popular name for henna at that time was its Middle Age Latin name "Alchanna" which lasted well into the 18th century.

HENNA BECOMES INTEGRAL TO INDIA'S MEDICINE

"Henna is a woman's life long friend."

—*Traditional Indian Proverb*

What many fail to understand is that while Ayurveda is the ancient form of healing in India, it wasn't always widely used there. Unani Tib (known as Unani medicine in India) which was brought to India by various invaders and forms of Tibetan and Chinese medicine ruled the land for ages. Vedic texts which provide the brunt of the information include 4 ancient books describing the healing method of Ayurvedic medicine which means "Science of Life" in Sanskrit. While touted as a real science, it greatly resembles the forms of healing in other locals including China and Egypt. Astrology and evil forces were still a serious culprit of disease and many will note the system of healing doesn't appear vary scientific compared to today's standards. It may have more complexities but it still can resemble astrology at times, as other forms of traditional healing did. One has the Chakra (meaning wheel in Sanskrit) system of energy points on the body and the Chinese had Acupuncture (meridians) and the

Egyptians, Reflexology. In Ayurveda one must balance the 3 Doshas and in Chinese medicine one must balance the five elements Wu-xing of nature and in Greek medicine one had to balance the four humors. Ayurveda also recognizes a life force in all called Prana and so does Chinese medicine calling it Qi and Egyptian medicine in the form of Ra. The similarities are very much there and shows how throughout the centuries they played off one another. While henna is included in areas of Ayurveda now and Sanskrit names, one being Madayantika, have been found, it doesn't seem to have been an important botanical like Neem is for instance. It also wasn't included in the original texts. In Unani medicine it is quite evident that it was and still is a frequently used medicinal plant. The majority of information pertaining to henna's use in India is extracted from Unani instead of true Ayurveda. Of course after henna was introduced to India, more and more uses were incorporated into the traditional healing techniques. Also it should be noted that Ayurveda is like Yoga, it is subject to many interpretations which has resulted in many branches forming which may or may not incorporate newer thinking. Many people attribute the main uses of henna in Ayurveda as:

—Hair health and baldness
—Asthma, coughing and breathing problems
—Colic pains
—Bleeding piles and skin problems
—Birth control
—Impotence
—Dysentery (amoeiasis)

According to Indian physicians, using certain portions of the henna plant for the given a condition was (and still is) quite important. As seen in the chemical makeup of the henna plant you can see how this was a good observation. The bark (which could be taken from older bushes) was used to sedate people and for its astringent qualities. It was also specifically used for jaundice, leprosy and enlargement of the spleen. The seeds (or dried fruit / berries) for use as a deodorant and for curing weakness. The flowers for cooling the body and helping with sleep. The leaves have many applications and were sometimes left on the twigs for use as a deodorant and in cleansing the body. Decoctions and infusions could be created from them which was intern used both internally and externally for various conditions, especially heat related. Due to the fact portions of India can get extremely hot, heat prostration is a real problem. Physicians early on were able to separate the various forms of over heating and their causes including Santaap (heat as a result of anger and frustration), Daah (heat and burning coming from inside the body) and Taap (heat resulting from outside sources such as the sun which was causing the skin to become hot and burned). Henna was used to cool and treat all of these forms of overheating. Also we must remember the essential oil or Attar of henna which is also employed in Ayurvedic medicine quite frequently, particularly in the Chakra department. Applying a healing essential oil to a specific Chakra was thought to heighten its ability to heal one and the corresponding organs to work properly. Most indication shows it being used for the Crown Chakra which sits over the head and is now associated with the pineal gland. It has also been said to help the heart Chakra as well as elevate pitta. Traditionally the oil was dabbed over where the Chakra was thought to originate or via the compress method. Just like Galen's theory, Ayurveda has humors which, when unbalanced, are thought to cause illness. The henna essential oil has been used for treating all three humors, vata, pitta and kapha (3 Doshas). Most likely henna is used on vata people the most as it is similar to the Melancholic type person from Galenical medicine. Further indication of this is the use of henna in attar form to heighten vata. Like Galenical medicine, Ayurveda uses a system of six tastes (Rasa). Henna is known as being a bitter or Tikta herb and an astringent or Kashaya herb. Bitter tasting herbs are thought to be comprised of the elements air and ether which intern heighten vata and reduces pitta and kapha. This is interesting because this can also be the reason henna is used to reduce fevers and cure skin disease as bitter herbs are thought to absorb phlegm and cleanse the body of "fire toxins", the cause of the aforementioned conditions. Just like Unani Tib medicine, henna is also thought of as being Shita or cold. Once again like the 6 tastes, herbs are sorted by their climate / energy called Virya as well (comprising of hot, dry, damp and cold botanicals). Ayurveda also has an area, like Avicenna's herbs effect on human physiology, where henna is noted as being Katu or pungent on the digestive system. Below are examples of some traditional Indian concoctions for healing specific conditions using henna. Although slated as being Ayurvedic, most have Unani influence and could very well be ancient Unani Tib recipes.

Headaches
—Combine henna leaves with the leaves of *Datura* and *Tamarindus*. To that add salt. This concoction is to be applied externally to the spine.

—Combine henna leaves with *Arrus precatorius* and *Tamarindus* with salt until it forms a paste. Then apply this to the soles of the feet.

—The flowers of henna are crushed and mixed with vinegar. The liquid is then applied to the forehead. This concoction is especially used when the headache is a result of heat.

Prickly Heat
—Place fresh henna leaves into water to make an infusion. This is then applied to effected areas. Fresh henna leaves do not stain like dried leaves do so this would not impart color to the skin, except perhaps a slight yellow tinge that would dissipate in a day.

Abdominal Pain
—The root of henna and the following should be ground together to form a paste; *Gandha, Ranga, Cocos micifera, Ita anla* and the tuber of *Moimorida dioica*. This paste is then formed into pellets, dried and taken internally for a prescribed number of days. (not safe)

Nausea
—The henna leaves are ground into a fine powder and used as snuff (Nasya) to relieve the nausea for a certain number of prescribed days. (not safe)

Fungal Foot Problems
—Fresh henna leaves are ground into a paste and lime juice is added. This is then applied to the feet to both prevent and treat fungal infections, especially Athletes foot, known in ancient times as "burning feet". This concoction could also be used for ringworm. In areas that are too poor to afford shoes, many times a thick paste of the above mixture is applied to feet to protect them from a whole host of harmful elements in the ground.

Skin Diseases
—The leaves of the fresh henna plant is combined with the rhizome of *Curccuma longa* to form a thick paste. The above is applied to the effected skin areas for about a week.

Heavy Bleeding (in women)
—The fresh leaves of henna are ground into a paste and pressed through a fine sieve (or piece of cheese cloth). The collected juice is then taken by the spoonfuls before eating a number of times a day an throughout a number of cycles. (not safe)

There are also many recipes for jaundice but since there is no real connection to henna helping one with such a condition, I will not bother to keep its use for such alive. Jaundice in an adult is a highly serious condition and needs real medical treatment. At times, the traditional recipes do more harm than good. The last concoction given for heavy bleeding in women is an intriguing one. Instead of helping the bleeding to stop, all indications are the bleeding would become worse, especially as the henna is taken internally. Women many times are the ones to get substandard medicine and that seems to be a dose of it right there. Other serious problems have come from Indian people trying to use henna on 2nd and 3rd degree burns. The non-sterile henna only made matters worse and hospital workers have to use more extreme measures to keep the person alive and combat infection. On the other hand a gel was formulated and tested on women in India, with very interesting findings. Just as the henna had inhibited the bacteria in the ground water, it was used for infections in the vaginal area in the form of a gel. Gels are easily created by combining aloe vera and liquefied or ground herbs and may be used in cases of both yeast and bacterial infections. Henna has long been thought as a women's life long friend in India and this can be seen by the many ways women employ it for both beauty and health. A specific helpful way henna has been used is as a birth control agent. Many traditional herbal texts from India tout it as being so but many have ignored this important effect of henna. First of all, texts conclude that henna has been used for such purposes both in India and Africa quite prevalently. We also know that henna contains a chemical called Beta-Sitosterol which acts as an anti-fertility agent in women and spermicide in men. Added to that is, one of the few ancient natural birth control's to be lab tested called Avrodhak (which was found to be henna) which actually showed some pronounced effects on fertility. When tested on lab animals, the henna prevented 60% of pregnancies without causing side effects such as weight loss or disrupted cycles. It did however continue to work 2 months after it was ceased to be given to rats whom received high dosages of it. Dosages that likely were under 100 mg's, as that amount and higher has been found to kill lab rats. The Avrodhak is just one of many ancient botanicals used in India for such birth control effects but certainly it is one of the few to actually work when lab tested. Henna's contraceptive effects have also been tested in China where a decoction of seeds / berries was made using both water and alcohol. This was in turn given to female lab mice and found to prevent pregnancy. Now before you start throwing away your henna for fear of being sterilized, remember the birth rate in India was around 5.3 and is now only dropping due to modern birth control products being introduced to the country. When used externally and occasionally, one should not suffer from the anti-fertility effects. However, if you are having fertility problems or are looking to conceive, I would suggest not using henna in any way whatsoever. This was you can be 100% certain that is played no roll in the outcome of your attempts.

Today in India, henna is being tested for all sorts of applications in the laboratory, including for cancer prevention. A good number of chemicals that makeup henna contain cancer fighting agents which researchers are looking into further through the use of laboratory animals. For a list read chapter 7. Ancient Egyptian people have been found to have rarely suffered or died from cancer and perhaps their constant use of henna contributed to this. Henna is also being tested for applications in fighting various fungal and bacterial conditions and even memory loss. According to one study, ether extracted from henna was given to mice who intern showed greater brain activity. Siddha medicine, which is another form of herbal / holistic medicine in India uses both henna leaves and berries / seeds (*L. alba*) to fight AIDS. No evidence has been presented that it has shown any promise but henna has been long used for various STD's throughout history. Instead of helping cure AIDS as a whole, henna is most likely more helpful to the conditions associated with AIDS which include skin cancers and candeda infections. Due in part to many doctoral thesis being done on henna, especially in India, many new positive health findings are being documented and hopefully in the coming years henna will be used to treat conditions both safely and successfully.

HENNA USE IN OTHER FORMS OF TRADITIONAL MEDICINE

"Finding henna is like finding a wild, centuries old Ginseng root."

—*Traditional Chinese Proverb*

While Chinese henna use may or may not predate Ayurvedic medicinal use, they too used henna both cosmetically and medicinally. Due to the spread of henna's use, it now naturally grows in areas of the Yellow River region and also quite prevalently in Southern China. China's use of henna is somewhat different than other areas as they use the gall (Imperial Gall) of the henna leaf (Wu-Bai-Zu / Wu-Pei-Tzu) and can be dated back to about 250 BCE. Firstly, a gall is an area of a plant where an insect borrowed down into the leaf or stem and the plant then attempted to recover from the damage caused. A large round bump usually occurs where the plant tissue has become inflamed. Due to the plant trying to mend itself, the galled area is usually more potent medicinally than normal areas of the same plant. These galls are then scraped off of the Chinese henna plant's leaf or stem for use in a number of concoctions. The henna plant was held in high esteem by the Chinese peoples and they likened it to "finding a wild, centuries old Ginseng root". Otherwise it was very expensive and hard to find. Surely henna was something only the rich could afford to use and obtain. Just like in other Asian areas, the henna plant was employed medicinally in China to treat headaches of all sorts. Many traditional formulations were also created from it including hair growth stimulation tonics. It was thought that the henna would cause the hair to grow stronger and more prevalently if henna was applied to the scalp. In addition to the galls, the Chinese used the berries / seeds of the henna plant and even the roots, flowers and leaves at times. Zhi-jia-hua, the traditional Chinese name for henna meaning "fingernail flower", is part of a group of Lawsone containing plants referred to as "Garden Balsam" or "Feng-xian". The prized plants were the ones with either white or red flowers as they were thought to contain higher amounts of Lawsone and tannins. What needs to be noted though is like the case of Europe having its own "true" red dyeing plant, Alkanet, so did the Chinese by way of their Garden Balsam (*Impatiens balsamina*). In India it has been documented that Chinese Garden Balsam was used as a henna substitute so obviously it died the skin well enough. This can vie for why henna use can appear more widespread in China then it really was. The henna however which was brought over during trade was found to be superior to Garden Balsam and its rarity made it more alluring to richer Chinese peoples. Traditionally henna was used medicinally for a number of conditions, many of which mirror those of other locals including Egypt and India. As with other forms of ancient medicine, traditional Chinese medicine or TCM for short, divides botanicals into various groups including by taste. According to ancient descriptions, henna falls under two of the five tastes, "bitter" (associated with Yin) and "pungent" (associated with Yang). This serves to tell us much about why henna was traditionally in China and for exactly what organs / functions. Following the network chart of the five elements (Wu-xing: fire, earth, metal, water and wood) we see that bitter herbs fall under the "sign" of "fire" and pungent under the sign of "metal". Bitter, fire herbs are associated with the season of summer (when henna flowers are traditionally picked in China), the emotions joy and mania, the color red (one of the prized henna flower colors) and parts of the body which include the heart, tongue and blood. It is considered to cool the body / mind and direct "qi" (life force) downward. This is most likely why henna was used in China to promote blood circulation, reduce swelling, stop pain, dispel toxins and so forth. This also is quite interesting because fire element people were thought to be prone to overheating in summer and botanicals such as henna was used to cool the body and calm the mind. It was also used to detoxify a condition known as Heart Fire which resulted in a flush face. Pungent, metal herbs on the other hand are associated with the season of fall (the other traditional harvesting time for henna, specifically berries / seeds in China), the emotion grief, the color white (the other prized henna flower color) and the parts of the body which include the skin and the nose. Using henna as a metal herb was thought to also help with the movement of blood in addition to curing rheumatism and colds. Just like with the Laws of Signatures, one can see how the use of flower color association, tastes, smells, etc., was incorporated in TCM to predict and decipher a botanicals effect on the body before using it medicinally. This also shows why henna was used for conditions such as whooping cough and "spitting blood" which isn't noted in other forms of medicine. According to TCM metal herbs are indicated for such lung conditions.

The henna is administered in a number of ways including internally. This is interesting because *externally*, henna is considered in TCM to be non-toxic. Internally on the other hand it is considered "slightly toxic in nature". Due to this, henna is given in small amounts internally, about 9 g. of dried henna or about 30 g. fresh. A strong decoction is made using water and is prescribed by a practitioner for a certain number of days or weeks. It also is at times used in pills or pellets. Henna flowers on the other hand would be sun dried and powdered. They are traditionally prescribed for a dosage of 1 to 3 g. dried or 3 g. fresh form. In TCM the flowers are considered to have the same medicinal qualities as the rest of the shrub or galls. Externally, the flowers and leaves are sometimes crushed into a paste and applied to the skin or rendered in water to create an external wash. The same is true for the roots which is prescribed for a dosage of 9 g. and the berries / seeds are given in 2 g. dosages. They can be powdered quite easily, when dried and applied externally for various conditions. According to Li Shi-zhen, a 16th century Chinese herbalist, berries / seeds were used to dislodge fish bones from the throat and caused them to become softer. He warns thought that the above should be thoroughly washed from the mouth in order to prevent tooth damage. Like the fish bones, a softening to the teeth was thought to occur with prolonged exposure. Many of the traditional recipes seem to indicate henna was both used singularly and in combinations with other traditional herbs. These include using the powdered flowers with honey to treat a lack of menstruation in women. A decoction combined with sugar was created to treat various types of coughs and a mixture of henna and wine to treat rheumatism and headaches.

Very similar to TCM is the herbal medicine practiced by Tibetan Buddhist monks and peoples. Starting around the 4th century AD monks started to research and combine the medical knowledge of other countries including Persia, China, Greece and India with their own pre-Buddhist Tibetan shamanic healing techniques. By the 11th century they had a well established form of medicine of their own which continued to evolve. In TTBM (traditional Tibetan Buddhist medicine), henna falls under the sign of "ground" or "earth" according to its astringent taste. Like TCM there are 5 elements which herbs can fall under called Mukhabuda and includes earth, water, fire, wind and wood or space. Henna's earth status indicates it would be used for afflictions such as cuts and wounds, fungal infections (such as that of the nails), diarrhea and any adverse condition, not surprisingly, related to hot summer months. Fresh henna flowers were also said to be incorporated into Buddhist gatherings for an aromatheraputic effect, especially around the traditional 3 month Lent. Another form of medicine TMC influenced was "kampo" which is the medicine of Japan. Henna was obviously introduced to Japan by China as it is called "tsume hana" or fingernail flower there as well. Likely this happened during the Tang period and was reserved for the very rich. In kempo, henna would also fall under the sign of Earth which is like TCM in that it is a part of the traditional 5 elements called "go-gyo". Henna was likely used in herbal formulations in much the same manner as in TCM.

The most common medicinal uses of henna in TCM include:
—Skin injuries and conditions such as sores, boils, carbuncles, wounds, etc.
—Menstruation problems (lack there of)
—Fungus and bacterial infections
—Blood and body purification
—Hair growth

Professional Herbalist Andrew Bently has studied the use of henna in Central Asia and has compiled his findings from his fieldwork for this book. Note how the Kazakh people follow what seems to be a form of the early "Greek Model" of four humors, as seen in traditional Unani medicine. This would make a lot of sense as one third of the population is Muslim. Ruled by the Mongols from the 13th to 18th century and then by Russia, the below also sheds light onto the speculation that henna body art was started or spread by Mongolian peoples as well.

> "Henna Grows on Rocks" is a proverb among the traditional nomadic Kazakh people of Central Asia. Like most folk wisdom, it can be understood in more than one way. It means that adversity builds character, that good can be found among the bad, and that anyone can become good, even if they don't come from a privileged background.

This proverb and its meaning demonstrate that henna is highly esteemed among the traditional peoples of its range. But why, exactly, is henna held in such regard? Partly, no doubt, for its use as a pigment. Henna bodyart is famous throughout the world. Among the Kazakh people, however, henna bodyart does not appear to be traditional. Henna is used in dying material and embroidery thread, but so are hundreds of other local plants. Another reason why rock-nurtured henna is regarded as a diamond in the rough by the people of this region is the fact that it is used in traditional medicine and hygiene.

In terms of Traditional Kazakh Medicine, it has a drawing effect on the element of earth (Zher) within the body. This effect could be predicted, the reasoning goes, by the fact that henna does grow among rocks, and must have a strong drawing power to be able to draw substance to itself from such hard soil. This also explains why other rock-grown plants such as many lichens have similar properties.

One of the most common uses for henna is as a deodorant. Traditional people in Kazakhstan, Uzbekistan, and other areas sometimes apply powdered henna to the underarms and the feet or keep leaves of the plant in the shoes, to prevent foul smell. This is also believed to protect against all sots of infections.

Henna leaf as a poultice or plaster is also used for arthritis. The plant is believed to draw the corruption which causes the arthritis from the body. It is well regarded for its marked ability to reduce swelling and pains in the joints.

The drawing power of henna also make it useful as a wound herb. Henna is mildly astringent, and thus able to speed the healing of wounds. It is also reckoned by its users to prevent or reverse infections of all sorts, as well as infestation by insects, which can be a real problem in the areas where henna is used.

Henna's earth-drawing power is regarded as useful for inflammations such as those resulting from stings and bites of poisonous creatures, from allergic reactions and other rashes, and even benign and cancerous growths on or near the skin.

The element of earth, in Kazakh thought, composes not only the dirt and corruption which may enter our bodies, but also lends to substance to the body itself. Because of this fact, henna is also used to stimulate contractions during birth, helping to draw the child's body out from the mother. For this it is either poulticed on the lower back, or taken in the form of an infusion (please note that henna is somewhat poisonous.).

Because henna has an earth-drawing power, it may cause inflammation, irritation, hives, and other allergy type disorders in those individuals in whom the element of earth which composes their body is granular, or not very cohesive; a result, in the traditional line of reasoning, of the earth-element being drawn from where it should have been. This particular "granular" derangement of the earth element mostly results from physiological or psychological weathering, i.e. stress."

Traditionally, the main healer of the household was the woman. She was a nurse maid to children and babies and her husband. Through passed down "wives tales", she could help ease the discomfort of fevers and other conditions. While men were the ones who traditionally recorded the remedies used, women, as nurses performed most of the administration. As a result, many folk remedies from various countries use henna to heal. One from Turkey suggests people with eczema create a concoction of eggplant which has been roasted over an open fire and mixed with henna powder. Allowed to cool, this mixture was then placed on the effected area and then covered with cloth and gauze. Another less appealing recipe from Turkey for curing headaches uses a mixture of henna with animal bile, placed on the forehead and allowed to sit for a few hours under a gauze wrap. A Persian recipe calls for henna to be mixed with suet and applied to nail infections or wounds to heal them. In Morocco henna leaves are steeped in milk and given as a purgative. An important area in women provided healthcare was during childbirth and midwifery. Henna was and still is in some countries an important medicine during and after the birthing process. This makes sense as henna contains a number of chemicals that have been proven to help quicken the labor process. Specifically by bringing on and stimulating contractions. At times the whole plant would be symbolically placed in the hair or on the stomach. Henna flowers would also adorn the birthing chair used by many Middle Eastern women. In more severe cases internal dosages was needed. It also has proven antiseptic qualities which many midwives would use in an attempt to sterilize implements, such as a knife to cut the umbilical cord and to stop bleeding when applied topically. This is still practiced in many areas of the African and Middle Eastern regions where access to modern medicine is many times greatly limited and home births are common. Midwives also will many times care or at least check in with the woman after giving birth during the "birth-bedding" time. In many Middle Eastern cultures, women are instructed to remain in bed for a number of days; 40 is the normal span of time in Turkey. During this time the woman must only breastfeed their child. Getting up to even take a bath is forbidden, less "bad luck" fall over the entire household. During this time the mother may wear henna though to regain her beauty and ability to have more children and after 6 months, if the child is female, she may have her tiny hands dipped in henna. Prior to the bedding period, some cultures, such as India, perform "Am" or "Ame" which is a purification process. Most likely this is done because some measure of blood is involved during birth and blood from women is thought to be polluting and defile others. Many times henna is indicated for use on the hands and feet of new mothers as part of these ceremonies.

If a woman has lived for a good number of years into old-age and has been used as a mid-wife or outside healer, she is many times given the status of shaman, crone or "wise woman". Unfortunately, in some areas, even being a "wise woman" is not enough to allow a female to gather herbs for use medicinally. This is true of a number of nomadic peoples including the Vaidus of India who are known as a band of traveling healers. Women are thought to be too lowly to collect herbs which need to be taken from the wild as they do not stay in one place to cultivate them long enough. Women are seen as polluting and even their shadow is not allowed to cast over the precious herbs including henna. Even actual wild-crafting is a ritual in itself. Herbs are usually picked to coincide with the time of the year known as Uttara naksatra (lunar phase) to the Vaidus. Prior to going in search of the herbs, the male gathers must take purification baths, record the names of the herbs they will search for and visit a temple in order to pray that they find them. Herbs are then wild-crafted and placed in separate bags. Later they are mixed together or properly prepared for storage. Due to the extreme labor it takes to gather the herbs, they are very carefully dispensed. Once again like other cultures that are somewhat cut off from the rest of society, they provide a glimpse into ancient herbology and wild-crafting traditions. The Vaidus in particular may be likened to ancient pre-Chinese Tibetan herbology practices. Not all ancient cultures, especially nomads, had the ability to use agriculture to grow henna and other medicinal herbs so finding and gathering them in nature was the only way. This was strenuous, time consuming and many times left to a few shamanic people who knew what to look for. Making the mistake of picking the wrong botanical could cost people their lives.

HENNA'S USE IN EUROPEAN AND NEW WORLD MEDICINE

> "...*There is a powder brought to us from Africa, they commonly call it Alchena. If we boil it in lye till it be colored, and annoint our hair with it, it will dye them red for many days, that is indelible...*"
>
> —John Baptist Porta (Giambattista della Porta), 1558

While many note that henna was used only in the latter 19th century and for hair-colorant in European countries, early notation by herbalists and other writers suggest otherwise. A fine example is that of Nicholas Culpeper (1653) who calls henna "Alkanna":

> "*Alkanna:* Privet hath a binding quality, helping ulcers in the mouth, is good against burnings and scaldings, cherishes the nerves and sinews; boil it in white wine to wash the mouth, and in hog's grease for burnings and scaldings."

Culpeper was known for frequently taking information from ancient texts and his notations here appear to be directly taken from the writings of Dioscorides and Galen. The above, small notation about henna should be compared to the notation Culpeper made about Alkanet found in chapter 12 of this book. Culpeper was very involved in the Law of Signatures and other astrological forms of medicine. Thus, he would dig up the mythical and astrological value of herbs that shows such links…with henna he puts forth none because he couldn't find any. Henna, while it was cultivated in England as a privet hedge or Mignonette tree in warmer areas, was expensive to obtain and most frequently used for hair coloring. As seen by notes made about it in 18th century chemical term books of England. Henna was seen in the Old World by English explorers as the perfect treatment for very bad body odors and sweating with John Fryer (17th century) writing henna was used "…to restrain sweating and the filthy smells proceeding therefore.". James Grey Jackson in 1820 agrees writing that henna "allays the violence of perspiration in the part to which it is applied and imparts a coolness." John Gill (1730) wrote in his white powdered wig that a good oil could be made from henna and it induces sleep. He also quotes one Dr. Shaw as saying…"This beautiful and odoriferous plant, if it is not annually cut and kept low, grows 10 or 12 feet high, putting out its little flowers in clusters, which yield the most grateful smell which may be likened to camphor.". Its main function became masking odors whiffing in from the filthy streets. It was also used as an aphrodisiac when rubbed, fresh, on the body. The Victorians also used it for a few true to henna medicinal applications including for the treatment of bruises, headaches and also pain in the limbs. It was predominately used in fresh form including what we would call today Aromatherapy. Alkanet could have been one of the reasons it was not used more often medicinally. This is seen by the words of the German author C. H. Ebermayer (1821) who stated Pharmacists can do without using henna roots (radix alcannae verae) to color oils and alcohol red. Instead he suggested they could use the less expensive buglos or Alkanet. Primarily chemists and others who dealt with botanicals focused on henna's hair coloring and use as a cosmetic. Whole leaves were boxed and sold though during the latter 19th century for healing purposes, especially in the US and later numerous henna powders were created including Hopkin's "White Henna Compound".

SAFETY ISSUES OF USING HENNA MEDICINALLY

People who should not use henna:
—Pregnant women
—Children and babies (under 6 years of age)
—Women on prescription birth control or hormone replacement drugs
—Both men and women actively dealing with fertility problems or undergoing treatment for such
—People with a G6PD-deficiency (anemia blood disorder)
—People with skin sensitivities &/or are prone to Allergic Rhinitis

People who should be very careful and under supervision:
—People with compromised lungs including asthmatics and those with bronchitis
—People with allergies to PPD and chemical sensitivities (not a factor with pure henna)

The lack of clinical testing causes many to jump to the conclusion henna is perfectly safe for medicinal use or they simply say "health risks are not recorded". In reality the health risks have been recorded but mainly in portions of the globe Western medicine looks down upon. This includes Africa where henna has been used to induce abortions for many years, which resulted in poisoning and deaths of many women. According to the EPA, henna use medicinally has also been linked to lead poisoning, specifically in the Middle East. While this comes as a result of henna being contaminated (lead poisoning *not* abortions which are caused by henna's many hormonal stimulating constituents), it is still a real problem. One tourist stated she got blood poisoning after having henna applied topically in Zanzibar Africa and African babies were also contracting tetanus from the application of henna medicinally to the skin. As with all botanicals, negative repercussions can happen when respect for the botanical is lost. Contaminated water and land is used, many times to cut costs. The same is true for using dangerous pesticides and unsafe processing procedures later. The henna is then used for folk medicine and body art, in a world that has become so much more complex. Hundreds of years ago, while yes one could have an interaction between herbs and foods, people were not taking prescription drugs as they are now. Added to this is the many chronic conditions people are now diagnosed with which can lead to adverse interactions. Dangerous pesticides were also not a problem. Simply because a botanical, which is thought to be natural, is used medicinally doesn't automatically antiquate concerns for side effects and drug interactions. As with a great number of medicinal herbs, henna has side effects. Side effects that for certain people out weigh its helpfulness medicinally. Case in point is that of very young children and babies. In addition to contracting tetanus after henna was placed on the umbilical cord stump, babies who had henna applied to portions of their bodies showed red blood cell hemolysis (hyperbilirubinemia). Their skin is so delicate and allows for transcutaneous absorption of the very oxidizing Lawsone. This in turn caused death in some cases. Especially susceptible are babies with an anemia called G6PD-deficiency; a rare blood disorder which effects many African descended peoples. Added to that is the fact small children, under six can quite easily build up allergies to henna due to its mild sensitizing effects. This is due to the fact henna contains Mannitol among other natural sensitizing chemicals. For the same reason babies should not be given certain foods (such as shell fish, strawberries, etc.) while growing up because of the chance they can build up an allergy to it, the same is true for henna. Once they are older teens or adults, they will then not be able to use henna again because an allergic reaction will manifest itself. According to studies this can include dermal allergies or the triggering of asthma &/or bronchitis. According to medicinal texts, the occurrence of allergic dermal reactions in older children and adults is almost nil when pure henna is used. In Islamic traditions, henna applied to children for no real reason (such as cosmetic henna body art wise) is not allowed. Also we can remember the writings of Avicenna in that henna was must suited for *adult* women. In India as well it was thought improper for unmarried girls to wear henna. This most likely has contributed to the low levels of reported adverse reactions suffered by children. There are plenty of safe herbs that cane be used for young children to reduce fevers and clear up minor skin conditions. Talk with your physician about these safer alternatives to henna. As for the safety of using henna body art on children under 6 years of age, I say "no it is not safe!". Especially with the constant contamination problems henna has today which only compounds matters. Never, ever give henna to children internally and always keep it out of their reach. If they have swallowed some visit your local hospital. A final note, remember that the FDA in the US has not found henna to be safe for use on the skin, it is only regulated to be used on the hair. So be safe and don't use henna on children under 6.

Another problem I have seen quite well documented, is asthma and bronchitis which has come as a result of having frequent contact with henna. This included hair beauticians and people working in mills. It isn't quite certain why people are being so adversely effected but many researchers seem to have come to the conclusion it is an *allergic* reaction. According to the "Health & Safety Sheets" put out on henna, it *is* considered an allergen and produces a fine yellow dust. People with such lung conditions should be mindful of this and probably completely avoid henna. I would also imagine the henna dust which all too often kicks up would have a negative effect on one as well. Perhaps this is why henna mill work was traditionally seen as a very lowly position to have. It would be quite advisable to wear a protective mask when working with larger quantities of henna powder. This way it can't enter your lungs and cause inflammation or irritation. One occasion, while pouring henna powder out of a plastic bag, large plumes of dust caused me to cough uncontrollably. The taste it left in my mouth was also absolutely awful; placing henna in the bitter six taste category was quite sage of the ancients. I later found that the government recommends that if the coughing does not cease in a short amount of time and any burning of the throat or mouth occurs from the henna dust, one should rush to the hospital right away. I did not have such burning and am still alive to write about it so I guess my experience was not that severe. To help keep this from happening, they suggest wearing a mouth and nose cover. Some mills also spray a bit of oil on the henna to moisten it and keep it from kicking up as much.

Women must be especially careful when attempting to use henna medicinally. Since henna can cause hormonal changes in women, it is important to think before combining it with hormone replacement therapies and birth control prescriptions. The henna, in theory, can cause adverse drug interactions. Henna contains the property Stigmasterol which causes induced ovulation in women. This can intern interact with your birth controls ability to function properly. The Stigmasterol also stimulates estrogen production and can aid in a number of problems including drug interactions with prescription drugs and even stimulation of some active cancers. Other properties in henna can cause induced bleeding and menstruation. This is why it is very important to talk with your doctor or better yet pharmacist about possible interactions. Many physicians are unaware of henna's "side effects" and showing them chapter 7 will help them to give you better suggestions about using it safely according to your personal circumstances.

Pregnant women also should be very, very careful using henna, especially in the early trimesters. While some American family physicians have stated henna is perfectly fine during pregnancy, they did not have access to the chemical breakdown of the botanical found here. Most books that contain such information run over $200 USD and are exclusively published in India. The American Herbal Products Assoc. has put out a very important book all physicians should have and they state "History of internal use as an abortifacient is recorded in Africa" and it should *not* be ingested. I have found further documentation of this, including a survey of young Nigerian women; 58% stated they attempted using *henna* for abortion purposes (further explained in chapter 13). Chinese scientists also noted that extraction's of henna made in alcohol or water caused uterus contractions in lab animals. Dr. Bala Subramaniam M. noted "Lawsone has well-documented Mutagenticity activity" and since it is henna's main, active ingredient this is quite important. Mutagen activity can mean birth defects can result from the given substance. In the midst of this, I think it would be very irresponsible for a medical physician to endorse or condone henna use during pregnancy. Red raspberry leaf isn't recommended and it only has uterus stimulating effects. Henna has uterus and ovulation stimulating effects, emmenagogue activity and Lawsone (10,000 ppm), a known abortifacient. The abortifacient chemicals in henna are one of the main reasons it is *not* approved for use on the skin by the FDA (a rule that stands since the 60's). When henna is used on the dead hair cells, not enough makes it to the scalp to do much harm. Taking henna internally is of course very unsafe. Once again, instead of simply visiting your health care provider and asking "is henna safe during pregnancy?", bring this book with you and show them this chapter and chapter 7. Especially the chemical breakdown of the henna plant. While you will note, I make these warnings under the header of "medicinal" use of henna, these side effects can also plague those practicing henna body art, especially in the case of pregnant women. Remember the recounting of herbalist Andrew Bently and his notation of Kazakh women applying henna externally to hasten the birthing process. The normal internal use of henna in India has been around 3 g. of henna dissolved in water daily (I add this is *not* safe). The normal amount of henna used to perform body art on the feet and hands has been reported as 50 to 100 g. and with an exposure time of 12 hours. So it is quite profound that when doing body art, one is exposed to far greater amounts of Lawsone and other inherent chemicals than suggested safe by the PDR and other health factions. To non-pregnant women, this may be fine. For those who say henna doesn't enter the blood stream, I ask how it is possible people with G6PD-deficiency had severe blood cell damage according to studies? The reason for this is that some of henna's natural chemicals are able to skin deep enough down into the skin that it makes contact with the underlying blood vessels. This intern allows it to be carried throughout the body. If this were not the case, people with G6PD-deficientcy wouldn't have had blood cell damage while using henna externally. It is also important to point out , non-deficient red blood cells also have been found to suffer damage from henna. I myself saw the effects of the cell damage when I accidentally allow the henna to enter an open wound on my hand; the cut and blood instantly turned black in color from becoming oxidized. The rest of my hand was stained its normal orange-henna color. Lawsone in henna has also been found to quickly enter the skin and bind with skin cells, preventing inflammation from irritants; otherwise a suppression of histamine. This is of course very helpful in cases of poison ivy and such but it also shows how quickly the Lawsone portion of henna enters the skin. Cells that would be receptive to the Lawsone (to start the antihistamine or immune system suppression effect) are found in the Prickle layer (also known as the Mucosum and Stratum spinosum) of the epidermis which is right over the Basal layer and blood capillaries. What many simply don't understand or don't want to understand is even though the hennotannic acid element of henna has a high molecular weight, it doesn't mean other inherent chemicals and EO (essential oil) which do not can't make its way into your body. I asked an expert, who teaches Advanced Molecular Genetics Lab and is a Ph.D. candidate for Biomedical Research, about if even though hennotanic acid has a high weight, is it possible for other inherent chemicals to make its way into the blood? He replied…"As far as absorption. YES! Absolutely and unequivocally." and provided a number of research papers regarding henna use externally and blood damage. According to the Health & Safety Sheets put out by the US government, under section "SAX Toxicity Evaluation THR", it states that henna *is* "Absorbed by the skin" and warns it can cause irritation and other problems. In the "Acute / Chronic Hazards" area, it states henna "may be absorbed through the skin". Also, the EO or essential oil of henna which is found in the leaves and flowers, gives it its distinctive scent and contains numerous concentrations of various inherent chemicals which can be absorbed into the skin extremely quickly. Many studies have found, people that applied essential oil to their skin later had it in their blood stream. Even when henna is dried and ground, the essential oil is present. When the EO thoroughly dissipates, it results in the henna becoming dull in color and depleted of scent, which is a sign of it being stale. So, what this means is the EO of henna is a separate route of entry into the body which needs to be considered.

It is always very important to take your babies health and safety into consideration before your own beautification or urges to try something new. Henna is perfect for use in the very last trimester or after your baby is born, as is done in Africa, the Middle East, India and other locals. During breastfeeding, it may even work as Hops does to promote more milk production. Most likely do to henna's abortive properties, Indian women heed the warnings of the Hindu goddess Lakshmi (goddess of luck and happiness), who exclaims henna on a pregnant women pollutes them, causing *harm* to their unborn baby and problems with the pregnancy. Some old myths do have an ounce of truth to them.

Another group of people which need to be especially careful is those with a rare blood condition called G6PD-deficientcy (glucose-6-phosphate dehydrogenase) and should avoid the use of henna altogether. Numerous studies have found that people with this condition suffered blood cell damage and hemolysis, which is their case is quite serious. While predominately African descended people suffer from this genetic condition, people from the Middle East regions also need to be careful. Cases of henna negatively effecting other blood conditions such as Diabetes has also been reported, so it is advisable to speak extensively with your physician prior to using henna if you suffer from a blood disorder.

Lastly, the issue of ingesting henna and thus being poisoned needs to be cleared up once and for all! Just as I stated in my previous book "Mehndi: Rediscovering Henna Body Art", poisoning *can* occur and henna is *not* safe for internal use. Various health literature states that henna is for "external use only" including the Botanical Safety Handbook" (CRC) and henna is included in the FDA/CFSAN's Poisonous Plant Bibliography. What intrigued me was the documented use of henna as a poison in Mexico and how henna is considered slightly toxic in TCM. Once again, according to the Health & Safety Data Sheets, which are used by chemists and others in research fields, under the "Toxicity" area, pure henna was given to lab mice in a concentrations of 100 mg's. Remember that this amount and more is many times used externally for full hand to elbow and foot to knee Mehndi designs. The result was 50% (LD50) of the mice died from henna effecting their hearts and intestine (interaperitoneal). The same was also seen in a study which stated "Vitro observations showed that Lawsone was capable of causing oxidative hemolysis in a dose dependent manner in rats.". Thus, henna can poison you when improperly used and most likely slowly over time as well. Another thing to remember is in the contraceptive tests, while it was noted high concentrations were given to lab animals to cause extended sterilization, it must have been under 100 mg's as they were not killed. It can be assumed much lower dosages can have an adverse effect on female fertility.

BE VARY CAREFUL AND READ LABELS WELL

After taking an internal weight loss supplement which continued henna (why they included it is beyond me because nowhere have I found henna described or used for weight loss, not even in India) the result was palpitations, Tachycardia (irregular heart best), Tremors of the hands and legs and insomnia. Someone else using the same product reported shakes, headache, hot flashes and feverish, trouble sleeping, resulting in a seizure and yet another person suffered a very bad stroke. Of course there was other ingredients such as the popular ma-huang but still it isn't clear what amount of henna they used and how much it contributed to the horrible side effects. However the henna was indicated as one of the herbs thought to have been the main ingredient and thus cause. In another case, a person had a stroke (cranial hemorrhaging) after taking part in a study which used a combination of herbs including henna (once again ma-huang was present in the blend). The age of the person was not given though. Yet another weight loss herbal blend, which this time did not contain ma-hung (but the isolate ephedra) caused similar negative results. This time the henna extract *and* powdered leaf was included in the supplement. While this is not a complete list (extracted from the US SN/AEMS records) and the complaints about both ma-huang and ephedra are far more numerous, one can't really say how much of a part the henna actually played. It should also be pointed out that pharmaceutical grade henna does not include PPD/other illegal colorants and no mention of that was found at the SN/AEMS. This I hope puts to rest the notion bugs, dirt and other adulterating substances are the real cause of adverse side effects. In reality it comes from taking henna internally without medical supervision (even with supervision as seen with the study incident). Be very careful when buying herbal products created in India and China in addition to diet blends. Read the labels very carefully and if there isn't a label, pass and buy another brand. Pregnant women should be especially mindful of the above advice.

Chapter Six

❀

Aphrodite's Love Potion?
The Aphrodisiac Called Henna

> *"From the clustered henna and from orange groves,*
> *That with such perfumes fill the breeze*
> *As Peris to their sister bear,*
> *When from the summit of some lofty tree*
> *She hangs encaged, the captive of the Dives"*
>
> —Thalaba 1800

To tell the truth, I wasn't even going to devote a chapter to the Aphrodisiac-History of henna; I thought I could get away with placing bits and pieces in other chapters but alas there simply was to much circulating that needed to be discussed. It deserves its own chapter and here it lies. Yet another window in which to view a side of the chameleon of a botanical known as henna. Henna has long been revered as an aphrodesical charm, inducer of love, especially with notation to its intoxicating scent. Its name is even frequently used as a pet name for women in many areas of the Old World. It appears in poetry quite often in regards to love and admiration, especially in the Middle East and India. King Solomon's Songs are a fine example of henna's scent and flowers linking to poetic and lyrical love (7:11):

> "Do come, O my dear one, let us go forth to the field; do let us lodge among the henna plants. Do let us rise early and go to the vineyards, that we may see whether the vine (henna) has sprouted, the blossom has burst open."

Once again we can also revert back to Cleopatra who had a profound love for henna essential oil / perfume (Cyprinum) and all things romantic. Unlike Helen of Troy, Cleopatra is constantly recorded by the ancients as a "plain looking woman" or "not very beautiful" and even "old looking". Being that this may have been the case, she really had to use the seducing power of scent, including the use of henna oil. Gaining a flare for overindulgence from the Roman's, Cleopatra used mass quantities of expensive Cyprinum on the sails of her Nile barges, on herself and in the form of Kyphi and Aegyptium concoctions. Obviously it worked, to a certain extent. Archeologists are now delving further into the way Egyptians, including Cleopatra, viewed love and romance. Through many poems found and murals with hidden meanings, a new picture of the importance of aromatics are being found. The Egyptians indulged in perfume and associated it with romantic encounters. Even an ancient story which became perhaps the bases for Cinderella hinges on the scent of a beautiful maidens hair, which was most likely dyed with henna. Anointing her body with Cyprinum, Cleopatras true intentions during Anthony's visit may be really revealed. One romantic henna introduction theory was of when Cleopatra went to meet Julius Caesar and later marry him, her servants followed bringing henna products and seeds. This is said to have brought henna to the Mediterranean. However one must remember the writings of Cyprinum and henna by Theophrastes hundreds of years before Cleopatra's birth, obviously showing that she was far from the first person to introduce the plant to the Mediterranean, especially Greece and later Rome. In addition you have Homer's mention of henna on Cyprus and so on. This is simply one of those myths that defies reasoning yet continues to surface as being part of henna's history because it's romantic. It is likely many Egyptian queens used henna perfumes prior to Cleopatra. Even other Cleopatra's.

In India, after the introduction of *Lawsonia inermis*, later poets frequently called henna "Kurabaka" in Sanskrit, and likened the yellow flowers to lightening. In the lovers tale of Gita Govida, Radha jealously pictures the absent Hari placing Kurabaka (henna) flowers in Gopis' flowing hair. One of the more notable poems comes from an Indian equivalent to Shakespeare called Kalidasa who wrote:

> "Kurabaka, a tree with five petalled flowers, blossoms in the Spring. The attractive smell of its flowers creates a yearning for the company of the opposite sex."

And here, a denatured poem translated from ancient Egyptian hieroglyphics by Amelia Ann Edwards, who substituted henna for another less poetic sounding flower. It still gives a glimpse into what may have actually existed but didn't survive into our modern times:

> "Oh, flower of henna!
> My heart stands still in thy presence.
> I have made mine eyes brilliant for thee with kohl.
> When I behold thee, I fly to thee, oh my beloved!
> Oh, lord of my heart, sweet is this hour.
> An hour passed with thee is worth an hour of eternity!"

Quite naturally many have linked henna's aphrodisiac appeal to Cyprus's main inhabitant, Aphrodiate / Venus. As you will remember from chapters past, henna flourished on the rich island which spurred Homer to name it Cyprus, meaning henna trees in Latin. While a number of other plants, such as Myrtle are "attributed" to the goddess of love, henna shows no real connection as no literary sources of the ancient time specify use in goddess worship or specifically for love. This could have been for a number of reasons including the great expense it took to extract essential oil and perfume from the flowers and/or henna was simply not exotic enough as it grew so abundantly on the island. Further spawning of the notion of an Aphrodite—henna connection could have also occurred from misinterpreting its Latin name Cyprus (Kyprus) to Cypris which actually is another name for the goddess. Even though there are mounds of ancient mythic stories from Greece and Rome, henna does not appear to infiltrate any of them. This can also be seen from the lack of such information by Culpeper compared to Alkanet. Henna being attributed to Venus seems to be a 20th century addition to its history.

One thing that is very apparent from old writings is that heavy weight is placed on henna's scent for its main aphrodeical effect. The use of henna in more of a cosmetic manner (staining on the hands, feet, etc.) and a connection to love does not appear until much later in history. Some authors such as Jogendra Saksena have suggested women were so feeble minded that they actually become entranced by their own henna designs and drifted into an erotic orgy of some sort. This was supposed to keep them occupied and content while their mates were away. I don't know about you, but I've never had such an experience. I personally find it rather demeaning to women. Cleopatra is a better example of how henna could have been used in many areas of the ancient Old World as an aphrodisiac. This is most likely from the fact many botanicals were administered in an oil or ointment form. Most oils had a very long shelf life, even in quite warm areas and in addition to this, could be applied to thin skinned areas of the body for a very quick, euphoric effect. This is seen by lubricant recipes for vigor (for men) from India which suggest using the oil of henna flowers and applying it with the rising and setting of the sun. The flower of henna, which produces a dark red essential oil, does not stain the body as the leaves do. However, any oil based preparation, especially where essential oils are present, has the ability to go deeply into the skin and make contact with the bloodstream. The inherent chemicals can then be circulated about the body, giving the desired effect faster. Henna contains predominately sedative actions to the Olfactory system as well as uplifting (anti-depressant) and calming qualities. This would make henna very comforting, especially to women. More evidence that flowers were revered comes from a very old book called "The Perfumed Garden" (925 AD) which was written by Nefzaoui. Like an Arabian version of the Karma-Sutra, the Perfumed Garden states:

> "They (women) also make use of the flower of henna which they call Faria and they macerate the same in water, until it turns yellow, and thus supply themselves with a beverage which has almost the same effect of camphor."

In this case Nefzaoui relates that camphor was used by women who were having fits of jealousy against rivals (most likely other wives in the family) or consumed too much food or wine. Men on the other hand would have more pleasure with their wife after drinking the concoction and thus henna flower tea was used for the same. Mohammed was also quoted as suggesting men use henna for sexual vigor and other medicinal effects. In many areas of India and the Middle East it has been long used as a treatment for impotence. However, it is rarely indicated for fertility and with good reason as henna can cause infertility in both men and women, either internally or when applied topically. It can not be more strongly stressed that the flowers of henna, most likely due to their extreme natural fragrance, have always been thought of aphrodisiacaly. This was then passed onto the entire plant, including the orange stain it could produce. Remember that in some areas, such as Egypt, women would actually make their henna designs using the flowers or use *real* henna perfumed oil to moisten the henna powder to form a paste. You may wonder, what is at the root of henna's universal aphrodisiac scent? Well, according to the author Havelock Ellis in the 1935 French book "Studies of Sexual Psychology" and specifically the chapter on "Odors and Perfumes", henna flowers (*Lawsonia inermis*) remarkably have an undertone which mimics human sperm. Yes, you read right. Just like some flowers mimic human pheromones or male musk. Ellis claimed this was the root of henna's scent being so amours, especially to women, noting, "I suppose that the odor of these flowers must have a more powerful sexual effect on women than on men". Ellis concluded that women who enjoyed henna, must have enjoyed being with their significant other in general. In addition, virgins would not pick up on henna's allure. Quoting Sonnini a great deal, Ellis also writes:

> "Bloch Sexual Life of Our Times notices that certain women are excited sexually by the odor of the flowers of the chestnut tree. It seems that one finds an example very remarkable and very significant of the same odor in the flowers of the henna (*Lawsonia inermis*) plant which is of a very widespread use in Muslim countries to color the nails and other parts of the body. Its flowers diffuse the odor more suave, said Sonnini in Egypt one century ago and the women like very much to carry them, to decorate their houses with them, to carry some to the baths and to scent their bosoms with it. They can not admit that Christian and

Jewish women share with them this privilege. If one crushes the flowers between the fingers, this odor prevails and is indeed, the only perceptible odor. It is not astonishing that such a delicious flower provided to Eastern poetry several charming features and several similar amours uses."

Ellis included that Charles Sigisbert Sonnini, while in Egypt (1717), stated that the use of henna flowers among women seemed similar to henna's mention in the "Songs of Solomon" in the Bible. Sonnini's notations of henna flower use is very similar to what Loring recounted many years later on his wondering's of Egypt (which can be found in chapter 4 in its entirety). He noted that fresh henna flowers remained a woman's constant companion and was frequently given as gifts and used to scent the bosom. It was obvious, in the manner women were described as using henna, that they were addicted to its use. Ellis also wrote that women noted that they would hold henna (and other flowers with sperm undertones) dear because out in the fresh air, amid these same flowers was the best places to have romantic encounters. Once again very reminiscent of the Songs of Solomon but also the manner in which the ancient Egyptians enjoyed encounters themselves. It would be easy to see why Mohammed considered henna paste (and the designs created) a type of perfume, as seen when he said "...do not wear perfume while you are in the state of ihram nor touch henna, because it is a perfume." Ellis wrote that "freshly mowed hay" had a scent many women found sperm like. If one remembers, henna essential oil, fresh flowers and perfume has a tenacious leafy and tea-like backnote (or undertone). These undertones are likely what lends henna its sperm attribution. Thus, it is not surprising henna's *leaves*, when dried, has an amplified "hay" and sperm aroma many women likely subconsciously gravitated to. Hence Mohammed's words. This would explain why, pretty universally, unmarried virgins and windows were not allowed to use henna, sometimes in any capacity. For the window, it may have been a cruel reminder of life she would never have again. For the unmarried girl, it may prompt promiscuity. This would also cast major light on the nagging question of why henna in general is so associated with women and femininity. Why men never equally exploited it and generally find its scent obnoxious. Why it is considered a woman's life long friend and comfort when her mate was away. The exploration into henna aromatically appears to be the missing link in tying together its history and specifically its use as an aphrodisiac. Further evidence of henna flowers being inherently aphrdisical is the Chinese view of henna as a plant falling under the category / sign of "metal", according to the five elements "Wu-xing". Metal flowers are associated with "Po", the part of the soul which reflects ones true animals desires and wants. Making it a perfect love potion.

HENNA USE IN INDIAN TANTRIC RITUALS

A number of people have attempted to link henna's use to ancient Tantric rituals in India. Similar to the Goddess cults of Rome and other Old World areas, there appears to be no linking of intrinsic henna use. The main focus of documentation which some say points to henna use in Tantra is in the form of ancient murals and fresco in India. Problems arise however to these theories when one takes the time to note no depiction's exist of green or black henna paste being applied to women. Instead, women are shown adorned and *applying* a red substance to their body including delicate skin areas. Obviously, the substance being applied is already in a red state and being painted on. Henna paste can be extremely irritating to delicate skin and it also makes little sense to prepare days in advance for the rituals. The true substances being depicted, which are backed up by many literary sources, include the ground spice Tumeric (*Curcuma longa*). Able to produce a variant in color, from yellow to bright red, it can be applied to any part of the body without irritation. It also stains quite well and stays put for a few days. When Alkanet is added or other red colorants, Turmeric turns extremely vibrant red or pink. It is also thought to be native to India. Turmeric, which in addition to Harida is known as Nisha and Rajani (remember Rajani is also used to denote henna *now* in parts of India), is quite prevalently found in extremely old Vedic texts, unlike henna which isn't even noted as a love inducer in the Karma-Sutra. It also is a name (or as the Indian people put it, attribute) used to denote the Tantric elephant head deity Ganesha. Meaning "the golden one", Harida-Ganesha is believed to have been a deity worshipped prior to the formation of ancient India. Used primarily for fertility and a *basis* for the Tantric rituals, it had many cults devoted to it in ancient times. The use of the Turmeric was extremely important in the rituals which included staining portions of the body, hands and feet yellow or red. This was thought to bring a happy marriage and many children, hence why it was frequently performed by women. It was also used to make one's muscles limber prior to yoga which was frequently a part of the Tantric rituals. Powders were the most popular way of causing these coloration's of the body and they were also used to cover the deity during offerings. This is once again why simply relying on paintings and frescos can produce a distorted reality of henna's real use. When you take this into consideration with the lack of historical recounting that henna was specifically used for such purposes in India, it leaves but a manufactured "20th" century myth. As a final note, one may remember India, most likely since its introduction, used henna as a "birth control" which some called Avrodhak. World it make sense for fertility minded Tantric cults to use anti-fertility botanicals as part of rituals?

HENNA AND MENSTRUATION

Yet another linkage to fertility some try to make, most likely spurred by Dr. Mahendra Bhanawat's henna—blood comparison and henna being to the ancients "the blood of the earth", is that henna is somehow related to menstruation in women and thus used for fertility. Keeping in mind that henna is actually anti-fertility in nature, this theory too falls flat as an explanation for why women make designs on their hands with henna or think of it in an aphrodisiacal manner. Firstly, the only connection I could find with henna and menstruation is the fact henna can "bring on" and stimulate a woman to bleed. This was most likely used internally after a woman gave birth to a child and then evolved to the stylized "decorate a woman's hands and feet with henna" during the bedding recovery period. In ancient times breast feeding was the only way to feed a newborn and baby. Many women find that breast feeding causes their periods to return in a delayed manner. This is because certain female hormones are suppressed. Doctors and researchers today are making linkages to reduced breast cancers and other adverse health conditions women can suffer, from breast feeding, do to the body having a rest from the constant bombardment of female hormones. Some women however may have wanted their periods to return right away, thinking they could continue having more children faster and thus used the henna for such purposes. Unfortunately, while the henna could stimulate menstruation, it also could prevent pregnancies without disrupting cycles. So, while the women's cycle would start again it's quite possible the henna prevented new conception. The use of henna to stain the nails &/or hands after birth as seen in the late 19th century could be a stylized version of such practices. It still however does not suggest henna body art revolved around menstruation. In many cultures around the world and since very ancient times, women were many times shunned during their menstruation. Special tents or huts would be constructed for them to actually live in. For many, this would give a rest from home duties and such. The main polluting factors women were thought to posses, by men, was a direct result of menstruation. This most likely resulted from a lack of explanation for why women bled (the correct explanation was not discovered until the 19th century), the fact blood could carry illnesses (or was thought to) and the very disturbing smell rotting blood has about it. Any likening between blood and henna would have instantly put a demise to its aphrodisiacal nature to men. While menstruation may mean fertility, it can also mean a lack of conception. Other then to bring on menstruation after birth, no real connection is made between it and henna. The statement that henna was "the blood of the earth" to the ancients also comes, seemingly not from an ancient source, but from the words of Homer Smith (author of "Man and His Gods"). This raises doubts that the ancients actually did think of it as the blood of the earth. Mr. Smith wrote about henna in the context of the forming ancient Egyptian valley:

> "Flowers blossomed in profusion, [including] the Egyptian privet, said to be the flower of paradise because the dye henna, made from its leaves and stalks, was red, the life giving color."

It's likely he was merging the thought Muslim's had about henna being in the Garden of Eden with the ramblings he was making about the forming Nile valley. Some Muslims believe henna to be a plant that originated in the Garden of Eden. Others (such as in Turkey) believe henna is the earth of heaven, which is considered a great paradisiacal garden as opposed to simply white clouds. This likely became twisted somewhere into the notion nondescript ancients saw it as the blood of the earth. Once again it is very hard to prove a fertility based use for something when it is anti-fertility in nature. Thus, it's better to look at what it actually *is* useful for and you are much more likely to find original uses and the basis for henna body art.

Chapter Seven

❁

From the Inside Out
The Pharmacognisy of Henna

> "Alcanna (Lawsonia inermis)—The dried and powdered shoots and laves were used as dye or, with suitable medium, a cosmetic."
> —18th Century British Chemical Terms

The botanical makeup of henna is quite interesting. Firstly, I would like to mention that henna as it is today and henna as it was originally in ancient times is much different. Harsh pesticides are used in many occasions and chemicals are added, namely the infamous PPD, to bring the dye deeper into the skin &/or hair. The soil and water used to grow the henna can many times be contaminated too. This has accounted for many scientists to only study henna growing in the wild, as the henna that was milled was so contaminated it thwarted their efforts and findings. PPD (para-phenylenediamine / 1,4-phenylenediamine) has become a "quote on quote" normal part of the henna powders makeup to many researchers which I find quite sad. Even though henna sellers bend over backwards to assure you there is no PPD in their products, they may not have proof there is not. These individuals import henna from popular growing countries including India, Egypt and Pakistan in large batches. Many times making the trip in a recycled oil drum on a ship. They then re-package it in pretty little boxes. Never doing pharmacological spot testing unless perhaps a number of people get hurt. This has become a normal occurrence and it really needs to stop before someone or many become extremely ill as a result and some sort of sanctions are placed on henna as a whole. This has happened in the past to other botanicals that were used incorrectly or contaminated. In countries such as India for example, there is no FDA (Food & Drug Assoc.) like the US has or other government agencies with the main goal of protecting the public from various unsafe products. There is also no laws mandating companies to clearly list all ingredients, including chemicals inherent either. So many people were becoming ill in India from lead contaminated powders used for Holli that the government had to put out a mass warning not to buy them and suggested using turmeric and other spices instead. Still, these products are allowed to be shipped all over the world and used by unknowing people. The majority of the products I have seen for hair color, henna body art (Mehndi) and so on have absolutely no ingredients label at all, not even "pure henna" or *Lawsonia inermis*. The Food & Drug Administration, Dept. of Health & Human Services put out some very clear information for henna companies to follow but more so for those producing henna for hair color. Still, it shows that henna is *expected* not to be pure and guidelines have been set for this which includes the following:

> "It shall *not* contain more than 10% percent of plant material from *Lawsonia alba Lam.* other than the leaf and petiole and shall be free from admixture with material from any other species of plant (adulteration). Moisture, not more than 10% percent. Total ash, no more than 15% percent.

Acid-insoluble ash, not more than 5% percent. Lead (as Pb), not more than 20% parts per million. Arsenic (as As), not more than 3% parts per million."

Henna, according to the FDA is also exempt from gaining certification according to section 721(c) as it doesn't pose special harm to the general public. It does however state that henna is to be used on the hair *only* and not on the eyebrows, eyelashes or the skin. It also points out PPD and other cheaper parts of the henna plant is illegal to place into the powder, here in the US anyway. Many also are unaware that the US FDA does not recognize henna body art (Mehndi) or the use of henna on the skin as a safe practice and henna is not indicated by them for such use. Being that this is the case, companies selling henna body art products under the statement they are "US FDA approved" are setting themselves up for investigation &/or law suits. It also should be noted that henna is completely banned in Germany for use on the skin. In "3 exp. 1" of Germanys cosmetic code laws, henna was apparently revoked from the list of safe cosmetic colorants or additives. The Lawsone content was noted as being the reason for this revocation which took place in 1986. While henna is still allowed to be used on the hair, people caught using, selling or applying henna on the skin face both steep fines and possible jail time. The FDA has essentially banned henna use on the skin since the 60's. Back in 1997, one of the larger of the two henna companies from Pakistan had their products detained by Customs due to adulteration of their henna for use in Mehndi henna body art. While it *didn't* state they were adding PPD, it did say it was detained due to illegal color additives. This spurred the FDA to write the following statement which they encouraged to be publicized (it of course was not):

"In April, 1997, LOS-DO examined two shipments of a hair color product, brand names Zarqa and Almas, for color additives. Neither product has directions for use. However, the labels for both products declare henna as the sole ingredient and depict designs on the hands and feet. The color additive regulation 21 CFR 73.2190 specifically allows for the safe use of henna in coloring the hair *only*. The regulation does not allow for the safe use of henna to make colored designs directly on the skin, including the hands and feet."

The FDA went on to say that all products from these companies would be seized by Customs and not allowed to be sold in the US (this report was updated 01/2000). However in 2000 many people were still able to find Zarqa and Almas products being sold in ethnic stores. An 11/01/2000 update shows how the FDA is cracking down on not only henna powders containing illegal color additives but also for henna products used exclusively for body art as well!

"The article [henna] is subject to refusal of admission pursuant to Section 801(a)(3) in that the article appears not to be a *hair dye* and it appears to bear or contain, for the purpose of decorating the skin, a color additive within the meaning of Section 721(a) [adulteration Section 601(e)]."

The "Important Notice" as the FDA put it, stated henna from the following firms may be detained without physical sampling because they have been found to be contaminated or bear improper ingredient labels in the past. They include the companies Babulal Brijbhushan (India), Harumal Gangaram & Company (India), Hash Pharma (India), Federal Exports International (India), Zaiqa Food Industries (Pakistan), M. Manzoor & Company (Pakistan) and Mehram Spice Industries (Pakistan), as well as Zarqa and Almas of Pakistan. Henna powder has been detained from the following countries, in the order of frequency;

—India (notably with the brands Mumtaz Henna, Raun Harman Inc.'s Colored Henna Pastes and Delta Exports Henna Kits all being detained for illegal color adulteration)
—Pakistan (notably with Joshina and Safi Henna products including pre-made paste, henna powder and kits being detained for illegal color adulteration)
—Sudan (notably products by Saleem Z. Al-Altawi Trading which also had products coming from India to the US detained)
—Canada

Also smaller Mehndi / henna body art companies obtain their henna from pretty much the same sources (such as the above detained henna exporters). They claim its the best from so and so village, but in reality they use brokers over seas to get them the product for cheap and repackage it. If this is allowed to continue, peoples rights to use henna freely may be taken away, at least in the US. One good side to this is that America, the UK, Canada and so forth are big importers of henna. If Customs repeatedly holds such contaminated products, it will most likely start to send a real message to these henna companies. Make it pure, or it won't sell. The problem however is much of the henna produced is consumed by the country of origin such as India and Pakistan where contaminated products are seen as being normal. Since reports from the EPA has found lead in Middle Eastern henna and the FDA blood poisons such as tetanus in henna, it really leaves no doubt that Americans are being exposed to the same dangers when they use henna for body art. This is why I highly recommend growing your own henna plants, especially if you live in an erred or sub-tropical area, they make great out door hedges. Instructions for this can be found in chapter 13 of this book. I certainly am, and many more are doing it as well in the form of a tub house plant. They [who grow henna] remarked the henna they got from dyeing the leaves and crushing it themselves was so superior they will never return to commercial henna again. The color for body art was rich oranges and browns. It makes sense, you are growing the plant organically (something *never* done commercially), on a small scale and with rich loom. Then you are powdering it from the dried leaves right away. Commercial henna can be stored for months before you get it. Hopefully some organic farm in New Mexico USA or in the surrounding areas will do everyone a favor and start producing "pure" henna. Something seriously needs to be done before contaminated henna ruins the reputation of henna body art (Mehndi) and the botanical itself. Already TV programs and major newspapers are warning people "henna" causes skin burning, scaring and so forth without pointing out contamination is the real culprit and not the *Lawsonia* itself. Let me reiterate that illegal colorants were found in red henna as well as so called "black henna" products. This is seen by FDA records of detaining a batch of "red henna powder" for illegal color adulteration from the exporter Mehran Spice Company. The constant bashing of black henna techniques has caused a false sense of security with red henna products. Serious contamination's can and do effect all henna, with the most popular being Salmonella, Tetanus bacteria, Lead and animal droppings which can spread all sorts of disease. Major adulterative substances, added to make the henna stretch, include corn (maze) and castorbean leaves.

Of course, if you don't want or can't grow your own, try to buy henna from herbal companies. They many times have a higher standard for quality and their henna is most likely being inspected at the shipping ports. Once again, if the grower and company know it will be used medicinally, they have no need to bleach it, add color and so forth and will most likely produce a purer product. All of the above problems with henna *today* is the main reason I do not recommend any one particular company selling henna products.

HENNA IN THE TEST TUBE

"C10H6O3"

— *Henna's Molecular Equation*

Henna (C10H6O3) is known by a number of molecular synonyms including its C.I. # of 75480, otherwise Natural Orange number 6 which is used for dying. Notice that is *natural* orange and not red. That is because henna dyes orange. In order to dye red additives and procedures are needed. Its molecular equation is also given as C16H16O5 in some circumstances. Without a doubt, the most important constituents of the henna plant (chemical name 2-hydroxy-1, 4-naphthoquinone, said to have been discovered by Timmasi in 1916) are:

—Lawsone
—Napthaquinone
—Tannins (polyphenols)

These chemicals are what create the staining and tanning qualities of henna. Starting with tannins, also known as tannic and gallotannic acid, is an organic substance found in a great number of plants. In this case, it is a brown colored resinoid substance, sometimes referred to as hennotannic acid (coined by Abd-el-Aziz Herraory at the turn of the century); which has a high molecule weight of 174.1556 and does not allow for crystallization. Henna has also been found to contain an olive green resin which is soluble in both alcohol and ether. Most tannic acid is either colorless or pale yellow. Tannin prevents things from decaying and is a mordant for dye. When acted upon by the enzyme polyphenol oxidase, tannin turns things a deep red to brown in color. This occurs when the henna dries and the leaf cells burst, allowing for a chemical reaction. Henna can contain anywhere from 5 to 10% galotannic acid. Lawsone (2-hydroxy-1,4-naphthaquinone) also arises from the drying of the henna and is key in creating temporary or permanent dying of materials. It is also henna's active ingredient and most henna contains anywhere from 0.5 to 1.5 % percent Lawsone (lawson). Hence why dried leaves / powder is used in body art designs. The staining substance napthaquinone is what causes the main emmenogogue and oxytocin actions inherent to henna and is also its active ingredient. Emmenogogue is what causes induced menstruation (bleeding) in women. It also has a stimulating effect on female hormones causing uterus contractions and egg release (ovulation). Napthaquinons are natural pigments which range from yellow to orange and in its most vibrant stage, red. Henna has been found not to be cyanogenic. Iridoids, saponins (sapogenins) and proanthocyanidins are also believed absent. Alluminium accumulation is not found although lead has been, hence why it is important henna is grown in organic soil. Under a microscope one will see henna's pigment forms yellow crystallized needles or prisms. This yellow color changes when the Lawsone is exposed to air and begins to oxidize. Yellow is typically the color henna will dye un-mordared cloth. What many don't understand (or don't want to understand) is many cosmetic books focused on henna's coloring qualities. Why henna's molecular equation is even based on the Lawsone molecule. The many other chemical constituents of henna were and are still not being discussed. Simply because Lawsone may have a very high molecular weight, which some say doesn't allow it to reach the blood stream, doesn't mean one of its other chemicals which has its own molecular weight could not. Each of these chemicals in themselves has an effect, sometimes beneficial and sometimes negative on the body. In certain circumstances the beneficial qualities of one inherent chemical in henna will eradicate the negative effects of another. In other circumstances however, certain chemical actions are compounded by a number of chemicals, such as the anti-fertility effects, which in turn will cause such actions on the body. Likewise, as seen in the use of specific portions of henna for various problems in Unani (and other folk) medicine, portions of henna can have greatly different chemical actions on the system. One of the main actions henna has is an anti-cancer effect. This certainly eradicates the carcinogenic qualities some of the chemicals pose. However, not if henna is smoked. Henna should never be smoked as the tannin and other acid constituents are highly carcinogenic. For *years* people selling marijuana in the Middle East would use henna to stretch the batch or as a full substation which left many extremely ill and in need of hospitalization. James Duke and his database of crop / medicinal plants hosted by the US government is by far the most up-to-date and inclusive list of henna's chemical breakdown. CRC also publishes a book by James Duke which includes a good amount of henna's qualities as well. James Duke is a superb authority on herbs and plants in general, taking the time to investigate the entire botanical, as a botanist should. Not simply dwelling on its coloring or pigment qualities. He certainly is a much better source than cosmetic history books, many of which use information from the 19th century. Here, using data from Duke's database as well as other sources, I will attempt to show the numerous chemicals henna contains *in addition* to the Lawsone. Each one has its *own* effect on the body and other living organisms which needs to be considered. It also will help to explain why the ancients used it for what they really did, i.e. headaches, cooling the body, sleep, burns, etc. Note that some of the chemicals listed have actions next to them. The chemicals that don't, are examples of chemicals in henna which have yet to be tested in the lab to ascertain what they can do. This is because some are unique to henna.

Henna's chemical breakdown:

Entire Plant
—3-BETA-Hydroxy-20-OXO-30-Norlupane
—Acacetin-7-O-Glucoside
—Hennosides
—Laxanthones
—Luteolin (diuretic)
—Lupeol
—Resedine (hemagglutin)
(The main actions of the chemicals found present in the entire plant include antifungal and some antibacterial actions. Many of the properties however are unknown.)

Flowers
—Essential Oil (EO)
—Alpha-Ionone (allergenic)
(The main actions of all of these chemicals include allergenic qualities when applied to the skin. Likely why the flowers never really caught on for use on the skin for body art. The flowers also contain many fragrant chemicals which are very hard to extract.)

Leaves
—1,2-Dihydroxy-4-Glucosylnapthalene (candidcide / antibacterial)
—Apigenin-4'-Glucoside
—1,4-Naphthaquinone
—Apigenin-7-O-Beta-Glucoside
—Beta-Sitosterol (antifertility / estrogenic)
—Coumarins
—Fraxetin (antibacterial / sunscreen)
—Esculetin (sunscreen / myrorelaxant)
—Mammitol (emetic / nephrotoxic)

—Gallatic Acid (anticancer / astringent)
—Hennosides
—Lawsone (abortifacient / oxytoxic)
—Luteolin-3'-Glucoside
—Luteolin-7-O-Glucoside
—Pentosans
—Quinone
—Resin
—Scopoletin
—Stigmasterol (sedative / ovulant)
—Tannin
—Chlorophyll
(The main [compounded] functions of all of these chemicals include anti-cancer agents, UV protection, spermacide, abortive, emmenogogue, anti-fungal, anti-bacterial, astringent, antiaemorrhagic.)

Bark
—3Beta-30-Dihydroxylup-20(29)-ENE
—205-3Beta-30-Dihydroxylupane
—Betulinic Acid (cytotoxic / anticarcinomic)
—Betulin (cytotoxic / antifeedent)
—Hennadiol
—N-Triacontyl-N-Tridecanoate
(The main functions include many anticancer actions in addition to antifungal and antibacterial properties.)

Seeds / Berries
—Behenic Acid (skin rejuvenator)
—Arachidic Acid
—Carbohydrates
—Fiber
—Water

—Fat
—Linoleic Acid (insecticide)
—Oleic Acid (perfume / irritant)
—Mucilge
—Palmitic Acid
—Protein
—Stearic Acid (perfume)

(The many types of acids contained in the berries / seeds shows why it was likely left alone by animals and people, as they were unpalatable. The seeds also have emmenogogue activity and have been used for such in the Middle East and elsewhere.)

Henna also contains ash, mannite, resins, essential oil, flavonoids, sterols and so forth. Its normal pH is that of 5 or 7 and it is 60% soluble in water, 95% in methanol and 40% in alcohol. It is also soluble in carbon tetrachloride and acetic acid but insoluble in both benzene and ether. It has a melting point of 195.5 centigrade and a fusion point of 196. Oxidation—reduction potential is 358mV. According to the Ames' test, henna as a whole was found to be non-mutagenic. This however has to do with testing on henna and other high tannic containing botanicals such as walnut shells for the possibility of cancer causing properties. Specifically when either is used on the skin. Henna of course would not pose such risks as it has actually long been used for skin cancers / tumors and it contains many anti-cancer chemicals. The Lawsone on the other hand was indicated as having well known mutagenic activity not related to cancer but instead psoriasis and is one of the reasons why henna should not be used during pregnancy if at all possible. Mutagens typically effect the unborn child in a negative manner, especially during the first trimesters.

BE CAREFUL TRAVELING WITH HENNA

"There is naught here but henna flowers."

—*Arabian Nights, Sir Richard Burton's Translation*

Few know that tests to detect the presence of marijuana created in the 70's couldn't distinguish between the real thing and henna! Henna was the only other botanical that caused the false positive. Thus, it would be likely smart to not travel with henna minced in small pieces. This can very closely resemble marijuana, even though the two substances don't smell the same. Likely, today better tests have been created to detect marijuana but one never knows the testing devices used in poorer countries.

Chapter Eight

❁

Black is Beautiful
The Truth and Myths About Black Henna

"*Her splendid caftan and her diamond's blaze;*
One spreads the henna;
One with sable dye,
Wakes the dim lustre of her languid eye."

—Lucy Aikin, 1810

Black henna today has become the black sheep of sorts in the world of henna body art. This is with good reason too as black henna products being sold today are many times adulterated with PPD (1,4-phenylenediamine) and is *very* unsafe to use on the skin. The so called black henna powder, which is notable by its true black color, is really nothing more then black henna hair dye from India commonly called "kalie mehndi". Most likely, out of a lack of having real henna, people resorted to using the black henna hair dye for body art and a new, dangerous trend may have started. Many point out however that Africa and Persia seem to have very long traditions of using a type of henna body art which involved black henna. Most notably, *Lawsonia alba* is generally the species known for producing black Mehndi in India. In the Middle East, black henna, which always has a stigma of being of lessor quality than red henna, was used on the soles of the feet. Since the chemical PPD was not created until the early 20th century, the mystery of black henna body art persists. I believe I have found its true origins, which I will put forth here. The first sightings of women using an apparently black substance on the hands in a patterned manner, or dip style on the finger tips and palms, is in Medieval Persia. Of course such representation comes in the form of artwork. And, of course with art, things are not always what they seem. However, in a number of drawings, including from the Spanish book "The Book of Chess, Dice and Boardgames" by Alfonso X Elsabio 1283, one sees Islamic women depicted *together*, one wearing a black substance on the hands and the other is wearing an orange substance. This would seem to put out the explanation the artisan didn't have an orange color to use in his drawings and was forced to use black instead. Such black designs have also been seen in artwork representing Persian and Indian women together, with the Persian women wearing black designs and the Indian women wearing normal henna orange. This is all very intriguing. It has been explained that the women were using natural sources of ammonia such as animal urine, lime, black sult or linseed oil to blacken the henna designs. These explanations come from a number of sources including Charles Sigisbert Sonnini from 1717. He wrote:

> "The practice of dyeing the fingers with henna is common in Syria, where certain women have a mind that their hands should present the sufficiently disagreeable mixture of black and white. The hands which the henna had first reddened become of shining black by rubbing them with a composition of lime, sulphate ammoniac and honey."

In Africa, black henna became very popular, where the henna was exposed to extreme heat, lamp sult and good old dangerous PPD. African women and anyone with darker skin tones however *don't* have to resort to such measures to obtain a very dark to black henna design. Remember henna has melanin stimulating qualities and will always pick up the undertones of the skin. I have seen many of my students at my college courses whom were of African descent obtain an extremely dark color using red henna powder (*Lawsonia inermis*) and the recipes I provide in my book "Mehndi: Rediscovering Henna Body Art". This included on the palms of their hands. I personally have even obtained a shockingly dark color (for me) using fresh henna powder from my plants. I really don't see why women would use such darkening tactics in Africa and such *unless* the henna they were using was so old and stale it wasn't able to color properly or they were getting boxes of black henna hair dye and trying to use that as a substitute for pure henna powder. This all appears like 20th century additions to the history of body art and not relevant to what was used long ago. That aside, while many Persians can be dark toned, the women depicted in the artwork are light skinned and they appear side by side, interchangeably using regular orange coloring and the black on the palms, fingers and backs of their hands. Suggesting the black would require a different technique or substance. Since ammonia doesn't work to darken henna on the tops of the hands, fingers, wrists, feet and ankles…the notion that it was used in Medieval times for such purposes does not keep with the numerous depiction's I have seen. In addition, literary sources don't seem to mention blackened henna on women of the time. Henna is always spoken in the same breath as the colors red, orange and yellow. Such as in a Medieval Persian poem which has the line…"A tree hath its hands, with henna, red dyed…". Never have I found any poetry or other writings making notations such as, her hands were dyed black with henna, or something in that manner. This continues to draw away from the notion henna was really being depicted in these Medieval paintings and drawings. That is when I started researching the use of "Gall Ink" and how it *may* be the true source of the black body art being portrayed.

Even today, names used to denote "black henna" in many parts of the Old World including the Middle East and India, is really denoting Blue Indigo and powdered ink. Likewise I have seen a number of pictures of designs drawn on the skin using *ink* which were explained to be henna designs, even though they were really not. This makes it extremely hard to use pictures, even photographs for documenting henna body art as inks and other dark substances are being used and called henna. Not only today, but long ago a number of explores documented women in Egypt and other Islamic countries such as Persia using Indian ink to make permanent tattoo designs on the skin. The author of the 1877 book "A Confederate Soldier in Egypt" writes:

> "...flowers pricked in blue on her chin, between her eyes, on her arms and on the back of her hands to imitate gloves."

He then went on to write how she had the palms and fingernails of her hands stained with henna, which gave them a dingy color of brown. So, apparently, the blue ink and henna was being used at the same time. The author also noted, in another area of his book, that he observed women pricking black designs on the backs of their hands with Indian ink and orange henna was used on the palms. Another author, J. W. McGarvey in his 1880 book "Lands of the Bible" wrote:

> "The universal custom among the Arab women of tattooing is observed. The tattoo marks are on the lips, the chin, the cheeks and the forehead; and they are see on the backs of the hands and on the wrists. Frequently the fingernails

and the palms of the hands are dyed with henna, which imparts to them an orange tinge. Tattooing is done by men whose profession it is and they frequently visited our camp to offer us their services."

Such documentation of women using ink, even in a permanent tattoo form, among Arabian people is frequently found. In a poem by Muhammad Din Tilai (c. 1800's), a line reads "You have set a star of green between your brows.". In a much earlier poem (c. 600 AD) called "The Tattooed Girl", Yazid Ebn Moavia writes:

> "Her hands are filled with what I lack,
> And on her arms are pictures,
> Looking like files of ants forsaking the battalions,
> Or hail inlaid by broken clouds on green lawns."

Also it is recorded by Shabeni in 1787 that "the Arabs tattoo their hands and arms.". It seems, quite apparent that this would render some pictures of women in Persia and other areas with black designs, as being indicative of tattooing or the application of ink and not henna. Designs confined to the face have been called "harquus" (or harquz) which appeared to use liquefied myrrh instead of ink. The indication of black and blue ink designs to other parts of the body could be permanent tattooing or simply topically applying ink which would stain the skin for a number of days. This is many times seen on very old dolls. On the backs of the hands and on the arms, their are various black designs. On the same hand, one may then see the fingers capped or the entire palm dyed with henna.

More intriguing is how the so called black henna body art seems to directly coincide to the creation and subsequent popularity of Gall Ink.

Gall Ink is believed to have been first created around 300 AD. No recipes have survived however from that early. The oldest artifact found was a slip of parchment paper from 600 AD Egypt which used Gall Ink for the lettering. The first recipe for Gall Ink shows up by the 12th century in a book by Theophilus. The main ingredients were tannic acid extracted from Oak Apples (the gall of the Oak tree) and ferrous sulphate (copperas, green vitriol or salmortis), which was found naturally in Spain's soil. Other additives included Gum Arabic to give the ink thickness and other colorants such as Blue Indigo and wine. By the time of Theophilus, Gall Inks were extremely popular all over the Middle East and Europe and the main producers were *women*. In Medieval texts, making ink was suggested as part of a woman's daily homemaker duties and would be a supplement to the household income. Since there were no printing presses yet, ink was produced as a small cottage industry by families, with the main users being religious scribes, artisans, music writers, poets and so forth. This would give women some of the first chances of seeing how well ink could be used for staining things, such as their own hands and fingers. I then got the idea of doing an experiment. Since henna has a good amount of tannic (henotannic) acid in it, I wondered what would happen if I used the Medieval recipe for Gall Ink, substituting henna for the Oak Apple powder. Taking dried henna powder (*Lawsonia inermis*) I added the liquid of black tea, making an extremely thin paste and allowed it to steep for over 6 hours. I then took ferrous sulphate (oh! the joys of having a pharmacist trained mother, all sorts of chemistry equipment is always at my disposal.) powder which I mixed into the henna with a bamboo stick. After a few minutes the henna began to turn black! Blacker and blacker it turned until it was the color of ink. I used the stick to draw a picture on white paper, it was amazing. I then went on to dye some cotton gauze black and gray. On the label of the ferrous sulphate it said it was only hazardous if ingested, so I decided to try the mixture on my skin. Everyone remarked, when I showed them the dried henna paste designs, how intense black the color was. I left the henna on over night. The next day I rubbed off the charcoal paste and there it was, a color of henna body art I have never seen naturally occurring on *me* before. It was black! Since *Lawsonia alba* is traditionally thought of as being able to produce a black color, perhaps it contains more tannic acid and thus reacts better to the addition of ferrous sulphate. Salts of iron would probably work just as well as it was traditionally used in henna dyes. I tried adding the ferrous sulphate to pre-made henna paste, which I knew had ingredients that weren't too kosher and found it failed to turn black. So perhaps the ferrous sulphate can also be used as a marker of henna being real or not. Did I discover the real Persian black henna technique? Would this explain the black designs on women's hands in Medieval Spain, as they were right in the area of large amounts of ferrous earth and ink production? I would like to think so but it could also mean women were simply using ink on their skin too. A recipe of antiquity from Persia to create a black eyebrow dye may confirm my experiments however. It combined Oak Apple, henna leaves, sulphate and burnt copper to create a black stain.

Using liquid ammonia to darken henna designs however is not only caustic but disgusting and should not continue to be perpetuated. The designs appear darker not so much from a chemical reaction but because the skin is being *burned* into a leather like state. Think of how dark red henna can be on the Turkish raw hide henna lamps and drums. Ammonia consumes water from tissue thus burning it. Stinging and inflammation is not needed in order to suffer the toxic effects on the skin. The US government notes that if you're skin is exposed to ammonia, for even the shortest amount of time, you should run and submerge the skin in water in an attempt to draw the ammonia out. This is the only way of stopping the burning. Never use liquid household ammonia on your skin in an attempt to make your henna designs turn black! This is not only foolish but unneeded. This is the 21st century, do women still need to harm themselves in the name of beauty? Likewise, don't buy black henna products being sold today. If the dried powder is already in a black state, then you know it is *fake* &/or has the toxic PPD or some other dangerous colorant added to it. The results of using such powders are also very dismal. I know because I took it upon myself to try some on my hand (which I will never do again). At first it was very black and defined but only after 2 days, the black color had disappeared and the real, faint orange color of the henna showed through. This is also a good marker for any henna, if you notice the deep color dissipates rather quickly, yet a light orange remains for some time, something was likely added to make the henna appear unnaturally darker. Blue Indigo is quite safe on the other hand and you can experiment with *Lawsonia alba*. Dr Bala Subramaniam M. recounted that in the lab he extracted *Lawsonia alba*'s black coloring ability and when he mixed it with regular red henna, it produced a black dye with green undertones. It's very likely few have ever really experienced true *Lawsonia alba* and thus have not been able to naturally get the very dark color. The best thing to do is grow both varieties of plants and compare!

PPD MAKES HENNA BLACK AND BROWN

This really seems like it should belong in the chapter on henna's darker side but in reality, denatured and thus dangerous henna products have nothing to do with pure henna. Thus, I will include it here, as it is also associated most with black henna powder and body art. It is a modern intrusion on not only the art of henna body art but its medicinal value as well. Numerous people have suffered major reactions from contaminated henna and reporting on major television and print media has begun.

They vaguely warn all henna can burn you and cause horrible allergic reactions. Those involved in henna body art rage back, its not all henna…its just black henna products which have PPD added to them! Who is doing the better disservice to the public? Unfortunately, it is the latter mentioned group. This is because *yes*, now in our modern world, all henna can cause a serious health risk and *no* black henna is not the only product that is adulterated. By bashing black henna (which I agree is called for today), many feel that all red henna products are safe and validated. This simply is not true. Henna isn't recognized as being safe for use on the skin for a reason by the US FDA and one may be its constant chemical adulteration. All henna, no matter what color it is supposed to dye the skin or hair can be adulterated with PPD. This is because PPD itself is *not* black. It also has no scent. It can be blended even in green, natural henna powder without a trace. It also not only aids in coloring items black but other colors as well such as brown and red. This is seen by the numerous synonyms used for the chemical 1,4-phenylenediamine which is commonly known as PPD today. They include:

—P-phenylenediamine
—P-aminoaniline
—PARA
—BASF URSOL D
—P-benzenediamine
—Benzofur D
—C.I 78080
—C. I. Developer 13
—C.I Oxidation Base 10
—P-diaminobenzene
—Benzofur black 41867
—Zoba black D
—Benzofur brown 41866

—Benzofur yellow
—Pelagol grey
—Santoflex I. C.

These are just half of the names. A good many locally used chemical names also exist. As noted in the previous chapter, the US FDA has already found red henna, henna paste and henna kits containing illegal color adulteration that was unsafe. So red henna is not immune to PPD contamination. PPD is extremely toxic when it comes in contact with the skin &/or is swallowed. Thus it is not surprising when people suffer awful open wounds after leaving the henna paste containing PPD on for hours. The most foolish development is when some people add PPD containing hair dye to pure henna powder to get a darker design. PPD shows mutagenic activity and attacks the liver and kidneys. Poisoning may not show up for up to 6 hours after exposure. Lab studies using 250 mg's of PPD on a number of animals caused skin problems after 24 hours. PPD was originally created for use on hair, which is dead and only for a matter of minutes. It was never intended for use on the skin and certainly not for hours. In the US it is illegal to add chemical colorants to any henna product because the henna itself is supposed to be the colorant. To get around this, companies in India are starting to abandon the name henna and use "Mehndi" or "myrtle" on the product instead. To sum things up, just assume all henna is contaminated with PPD. BODD reports that the majority of cases concerning adverse skin reactions is from henna adulterated with PPD. It's foolish to suggest reading ingredient labels because the only legal additive to henna is *nothing* in the US. Other countries do not mandate that real ingredients be listed (or if they do, it is not enforced). Would people actually buy what they think is red henna if it had PPD combined with it? The only exception may be hair color which is *never* suitable to be used on the skin. The main reason PPD is added is to try and hide the fact the henna is old and stale. My advice is to go buy a henna plant and save yourself a trip to the hospital.

Symptoms of PPD Adulterated Henna Poisoning
—Weakening of the muscles
—Nausea and vomiting
—Unable to wake up (coma)
—Decreased body temperature (abnormally low body temperature)
—Pulmonary edema
—Pulse rate increases without the fall of blood pressure

If you are suffering from any of the above symptoms or have oozing sores where the henna paste was once applied, you need to be rushed to the hospital right away! According to a report by Kamil A. Abdullah in the Lancet, a 40 year old Saudi Arabian woman in 1993 collapsed after using henna on her feet and one hand. She was never able to finish decorating her other hand before passing out and being rushed to the hospital. Although PPD was not specifically determined to be the culprit, the neighbor who gave the woman the paste refused to explain what was used in it and the symptoms seemed classic for such a poisoning. The worst part is poisoning can happen over time, as is seen by a number of women being crippled because of toxic henna products they repeatedly used in the Middle East. Cyanosis can happen when PPD is absorbed into the skin and forms methemoglobin. It can also cause birth defects and permanent skin scarring. Another thing to remember is PPD is just one of a number of dangerous substances which finds its way into many henna powders. For more on other dangerous contaminants, read chapter 7.

Chapter Nine

❀

A Small Liberation for Women
The Mystery of Mehndi

"The nails and the palms of the hands are stained with henna cultivated there: The Arabs tattoo their hands and arms but not the people of Timbuktu."
—Shabeni, Descritption of Timbuktu 1787

The Many Names of Mehndi:
—Mehndi (most popular from Hindi)
—Mehedi
—Mendi
—Mendeed
—Mendhi
—Medi
—Marithond
—Maruthani
—Mehadi
—Mayilainanti

—Menhadi
—Monjuati
—Maidhakku
—Menhada
—Mailanci
—Mayilandi' gorante
—Mindi' bind
—Mayilanchi
—Mayilanji
—Marathondi
—Mehari
—Maratoli
—Mandana
—Mendy
—Meti
—Mandee
—Madaranga
—Medudi
—Muhanoni
—Mahandi
—Mayhendhie
—Muheni
—Maritondo
—Mayendi
—Marudani
—Mendie

The simple question many have neglected to answer *is*... "what does the word Mehndi mean?. If it is answered, a quick "the use of henna as body art" is the norm. While this is somewhat true *today*, Mehndi has a much older and truer meaning which has nothing to do with body art or henna. Mehndi really means "Myrtle" in Hindi and Sanskrit. Mehndi was usually teamed up with other words to specialize the type of myrtle being referred to (i.e. Vilayiti-Mehndi, etc.). Henna, when it was introduced to India, was thought of as a type of myrtle and thus was called "Hina-Mehndi". The "hina" is obviously a modification of the Arabic Hinna. However, over the years the hina was dropped and "Mehndi" alone started to be used. This is why we see Mehndi perfume / oil, Mehndi flowers and so forth. This is not a mistake of writers, as Mehndi does not truly mean "body art" using henna. In reality there is *no* very old word that means henna body art in India and no distinction is made between henna's uses, suggesting one or both of two things.

1.) Body art using henna was not an important, deliberate act but instead a result of using henna for other purposes, such as to cool the body or color the hair.
2.) Body art using henna, specifically in patterns and deliberately is a much newer addition to henna's history, perhaps first starting in the Middle Ages.

So, if one berates you about "Mehndi" not denoting the henna plant but instead the body art of staining the skin with henna, you can smile with the correct knowledge and utter a simple "I must disagree". Some have insinuated other prior meanings for Mehndi including "religious guide" and "Tumeric". I have never found any of these meanings mentioned in credible Indian Lexicons or dictionaries. These proposed meanings seem to be in reality an attempt to link henna body art in India to deeply esoteric or religious uses by women. One of the stranger portions of henna's history, especially in India is how one *type* of henna is always singled out as being better. It also has a different, un-Mehndi related name. "Hina Mehndi" (or simply hinna / menhada) is usually henna that is looked down upon as being inferior for creating body art. It is described as having broader leaves and very fragrant flowers. Thus it was said to be used more for perfume. The other type of henna is called Rajani (or Rajni) which has small leaves and better coloring ability. Since all the way back to the 18th century AD botanists have documented 2 species of henna growing in India (*L. inermis* and *L. alba*), these two species must be the ones being mentioned. Which are which though? *Lawsonia alba*, without being associated with Rajani, is frequently noted as being lowly compared to red henna (*Lawsonia inermis*). So perhaps *Lawsonia alba* is what is considered the true Hina Mehndi in India? It's all very confusing when local Hindi usage's are practiced to separate the two types. This however doesn't seem correct as Mehndi and Hinna (heena) are the most popular words for henna in India. Why would the inferior henna type become the most popular? Perhaps this is simply a way of slighting the Muslims who brought "Hinna" to India. As if to say…"we already had a better henna growing here before the Moguls", thus giving it an Indian name of Rajani. Maybe it's even a reversal of what women really thought about the two henna types. Rajani actually means "night" in Sanskrit and is usually teamed up with another word to denote a night blooming flower type. Henna is noted as smelling particularly sweet at night so this sounds plausible. It must have been originally teamed up with another real henna meaning word, as Rajani alone is somewhat ambiguous. Rajani I also just noticed (after looking through my notes) is used in Sanskrit alone and not in conjunction with other words to mean Turmeric! Oh, what a tangled web. This would certainly mean that Rajani was growing in India prior to the Moguls. What ever the case, it would suggest that *perhaps* the two species came at slightly different times *and* that henna was certainly introduced. As India repeatedly keeps referring to henna as a type of botanical that was both native and cosmetically used there (i.e. crape myrtle and turmeric). This may explain why henna appeared to be used only for hair color until the 12th century in India. This would also put to rest theories that henna as body art had a parallel beginning in India with no influence from other cultures. Using *other* botanicals for body art does not equal henna body art. The only reason henna body art is what it is, *is* because henna is being used. If Indian women were using turmeric then it would have been turmeric body art or body painting. Which certainly is the picture emerging when you begin looking at the linguistics of Indian names for henna. Also when you think back to the Ajanta and Ellora caves, which are likely fine examples of turmeric body art / painting. I have a very strong feeling that before the two species of henna (i.e. *L. inermis* and *L. alba*) were documented, Indian women were making a comparison between the introduced henna plant (Hina-Mehndi) and their native turmeric (*Curcuma longa Linn.* Rajani). They were probably trying to use the henna in fresh form (which is seen in very antique Indian recipes) and in a painting manner which of course produced dismal, yellow results compared to the instant red color achieved with their native turmeric or other plants. In fact, the first notation in India of women using henna on the nails (12th century) notes that a *yellow*, not red, color was achieved.

While many wish to tie esoteric connections to henna body art, there is really no evidence, especially ancient, to do so. Instead there is much data suggesting henna body art was really a simple result of women *and* men trying to cool their bodies from the extreme heat conditions of their surroundings. Over and over again one sees the same references to henna being applied to the hands and feet to help cool one. Then, even the designs or color left on the skin after the henna paste was removed was thought to keep one cool. Such notations are seen all over India, the Middle East, Africa and so on. It also makes more sense then other theories because henna does have cooling qualities to it. One has to remember in ancient times there was no air-conditioning so people would have greatly appreciated *any* little bit of cooling something could give. Henna is very unique in the way the paste seems to draw heat out of the body. In fact, in many areas including India, it was thought the more heat you had stored up in your body, the darker the henna designs would become. Men and women would frequently coat the soles of their feet with henna paste so that they would not suffer burns from the hot sand. Henna was also thought to strengthen any area in which it was applied, allowing one to work harder with the hands or feet. Pearl divers, fishermen and musicians would apply it for such purposes. It was also thought to keep away foul body odors. Holding a ball of henna paste in the hands was very popular and the majority of older depiction's of women with an orange color on their hands, have it confined to their fingers and palms. As if the color was made via holding something in the hand. In fact in many areas, including the Middle East and India, women would note that in the *olden days*, they would take a clump of henna paste, place it in their hands and curl their fingers around it. After that they would wrap some cloth around their hands and twine to keep it in place. In the morning they would remove the wrappings and henna to find an interesting design made from the henna failing to color the creases of the palm. This is also seen in some Indian weddings where the bride and groom hold henna between each others hands during the ceremony. This is over and above the designs already made on the brides hands, perhaps hearkening back to what was traditionally done before elaborate designs were created on the hands in India with henna. I personally believe from various sorts of evidence that henna body art progressed in the following manner:

1.) Henna color was applied to the skin as a result of coloring the hair or holding a ball / clump of henna in the hands to cool oneself.

2.) Henna paste was rolled in thin threads (called Jali in India) to form actual designs on the skin. This, early on was likely similar to the ropes of clay used to make pottery but later, became thread thin in order to create more intricate designs.

3.) Henna became more liquefied and applied with a twig or stick. Later when the English infiltrated India, they brought cake decorators which is today called a "cone" and is used to apply a very liquefied henna to the skin.

Some modern authors explain henna designs began in India when a large circle in the middle of the palm had smaller circles placed around it. These designs are said to be seen on old religious statues. However there is little indication that this is the beginning of henna designs in India, as Persian women of the same era had elaborate henna designs and frequently visited India. In addition these circles were not henna designs but instead "bindi" markings, which are explained further in this chapter. This is why circles were always present and never seemed to progress further. As part of the Solar Sringar (which is an extremely old form of beauty in India) women world draw elaborate floral designs on their face. So it makes little sense to not be as advanced with Mehndi designs as well, if the henna was really used since antiquity there (which it was not). Women have of course been making designs on their bodies for thousands of years using all sorts of substances. What makes henna unique is the manner in which it is applied. Instead of a pigment, the henna paste needs to sit on the skin for a long period of time and then is removed, creating a stained in design. It is however hard to tell using art if the designs sometimes seen on women's hands are really henna. One has to remember the artisan was likely a man, who could have been adding his own artistic interpretations instead of what was really being done by women of the time. Likewise, henna can change to any design idea of its user. There is no such thing as a "henna type" design. The only way that henna body art is unique is the actual use of henna. Women in India for an example have been using all sorts of colored substances to create body art prior to the introduction of henna. This is seen by the numerous names used to denote henna which have duel or prior meanings of other unrelated, coloring botanicals. This includes the native Indian crape myrtle, rue, baleria and so forth. One can also look at some of the old henna paste recipes from India to see how women were using other substances prior to henna's introduction. Specifically when fresh henna leaves are mixed with alkanet, catechu (*Acacia catechu*) and other intensely red and brown botanicals. Fresh leaves do not impart any color remotely related to red, so the addition of the other red botanicals is what is *really* creating the body art designs. I know this because I created a fresh henna paste neglecting the red and brown botanicals traditionally used in India and received only a faint, yellow stain on my palms which disappeared by the next day. This only makes sense because it is the drying of the henna which starts the chemical reactions needed to color in an orange manner. Similar to how tea leaves much be dried before you can use them. Today, henna appears to be the be-all, end-all use in India for body art but in reality, Indian women have been using Alkanet and even Garden Balsam for years to create designs on the hands, feet and other areas of the body. Professor Guatama Vajracharya from the University of Madison recounted in a Cardinal article by Carrie May Poniewaz that prior to the introduction of henna , Hindu, Buddhist and other Indian women used a dye from the Laksa tree to stain their hands and the bottoms of their feet. Specifically the fruits were used which was sap that would form hard jewel like pieces on the ground and impart an intense red color. Professor Vajracharya then went on to say that when henna was introduced by Muslims it quickly replaced Laksa because the henna plant was easier to find and grow. In order to get pieces of Laksa one had to forage in the forests of India while the henna could simply be cultivated. Likely the professor, like myself, noted that there are no depiction's of women applying a dark green or black substance in various old art. The substance is always, clearly already *red* in color. This would also pin point the fact henna does not have real *ancient* religious significance in India, thus explaining why so many

Indian people in general use henna. Not just Hindu peoples. While henna was important to Muslim peoples, it most likely was not to Indian peoples when it was first introduced. This is seen in how it was used for hair colorants and was neglected to be mentioned for anything Tantric related. To try and counter this fact, many Indian people are trying to place heavy religious weight on the designs used in Mehndi thus trying to carry that religious significance to henna as a whole. This of course is silly because henna has no set rules or set designs. Women would make the same designs on their skin (and have) using other substances as well as on the walls of their homes and on the floor. Henna's religious significance in India is a manufactured byproduct of the 1950's and its bloody independence. It has since been a tool of hate and racial / religious intolerance. One example comes from a 1940's book on the fighting among Muslim, Hindu and Sikh peoples. It stated that a group of Muslim men were tauntingly sent Mehndi (henna powder) and women's bangles, implying a lack of manliness. This resulted in the Muslims setting fire to Sihk and Hindu towns in Punjab. Other violence broke out including senseless stabbings and looting by Muslim as a result of the henna. Thus, many Hindu people even today taunt Muslims by saying henna is a part of their pagan religion and if they use it, they are going against what Islam teaches. Unfortunately, this has caused a number of people today to feel guilty about using henna, even though it may really be a part of their direct culture. Personally, I have surprisingly found people *living* in India to be much more realistic about henna's origins, explaining it originated in North Africa and was a gift from the Moguls or Egyptians. It's the people whom are removed from the country and living elsewhere that are absolutely belligerent about henna at times. There is no reasoning with them. Perhaps this is do to their feeling that their culture is slipping away in their alien surroundings. Henna, being important today in weddings, is one of the few items they have left to make them feel close to home. People however should ignore such behavior and use henna as they please, especially if one is Moroccan, African, Egyptian, Persian, Turkish, Greek, Spanish or any one of the many other ethnic groups which used henna not just for body art but also as a perfume or medicine. It truly may be a part of *your* culture and you, even as a general woman, should be able to use henna to bring you back to your own personal roots. When you think about it, henna body art is a fully woman created artform that would allow a women, many times living in a very restrained atmosphere, the ability to express themselves. The designs would sometimes be extremely intricate and sometimes rustic but like many artistic endeavors by women, was simply tossed into the realm of folkart. Using henna on the hands to create designs is truly liberating and a true artform. After the designs have set, they can be appreciated by many, even if one has to always wear a veil. Henna body art is not suppressing as some try to make it sound. Many have misconstrued the traditional privileges some Indian brides are given as an example of how suppressed henna *makes* a women. This is of course ridiculous and the result of those whom have never personally used henna trying to add a negative feminist twist. In reality, henna has long been used by many cultures and *midwives* as a suture like glue. The reason for this is because when henna paste dries, even if it is only moistened with water and nothing else, it adheres remarkably strong to anything it is applied to. This includes the skin! It's true at first, when you apply the henna paste in its wet form it can rub off…but who really cares. One simply applies a touch up. It dries however very fast and when it does, you don't even need to cover it in order to protect it from rubbing off. I have gone throughout a whole day of normal work and chores with hands or feet decorated with henna paste. Needless to say I was *not* incapacitated. Women in Yemen would many times wear the henna paste as the actual body art designs and I have done experiments to see how long henna paste could cling to the skin. Even with the rubbing of clothing, it held tight for over 5 days. The only reason women wrap henna designs is to keep them moister which may aid in better color outcome. Some Indian brides are allowed to forgo any housework for as long as their henna designs are apparent as it is thought to be a marker of how happy the marriage will be. Think of all the superstitious things brides in the US do to ensure a happy marriage. So it is not a good example of how henna is used on a regular basis. Many women simply use a twig or match stick to make little designs on their hands such as they would apply eye kohl and their normal jewelry. The sad part however today is the how many of the henna products women use are contaminated with dangerous substances and adulterants which greatly threatens the art form.

Some of the popularly theorized origins of henna body art include Morocco and the area which comprises modern Turkey. Unfortunately since the Berber's of Morocco had no form of writing, one is hard pressed to be able to definitively document and present such a theory. The case is also not helped that they collectively converted to Islam which has much religious weight placed on the use of henna. If they were Christians and using henna, it would have been more convincing to the theory of their starting the traditions. Likewise, henna body art starting in the Mediterranean or Near East is relatively unfounded, especially as there are no Akkadian names used to denote henna. This is what led to the use of the Persian Hinna, transformed to Kina, instead of an anciently used name in the area (as a comparison try tracing myrrh). If henna was so intrinsic to the area, why adopt a word from Persia or Arabic? John Fryer also writes in his 1681 letters (later compiled into the book "Fryer's East India and Persia") that Turks in Persia would import, from Africa, henna which was called Alkenna. It was used to stain their hair, nails and skin a red color. Why import something that your area is supposed to be known for? Obviously henna wasn't growing as abundantly as needed to supply the henna powder. The same is seen today where many countries consume all of the henna grown in their area and thus have to import it themselves to keep up with demand. It would make sense food crops would be grown on much of the land and henna was grown on any remaining areas. Such staining was explained by Fryer to cool the liver and keep away foul odors from the body. No mention of Evil Eye protection or the like was made, so it gives more credence that even in the 17th century, henna used in body art had an underlying medicinal factor attached to it. Such a cooling and medicinal factor, which is seen over and over again throughout history, is most likely the true basis for henna body art. Thus one has to ponder when do stains left on the hands or feet from using henna medicinally become recognized as body art? Looking at older depiction's, many will see a lack of ornamentation and only a complete staining of the palm &/or fingertips. Just as one would obtain from using henna in a medicinal manner. Since mummies have been reportedly found with similar staining, why is it so foreign to think they were the ones technically to start henna body art traditions? Permanent tattooing in general has very ancient roots in Egypt. Small female clay figures were found dating from 4000 BCE in Egypt with dots and dashes on their bodies. This however was not enough to warrant making the assumption tattooing was practiced in ancient Egypt. Then, when archeologists started looking closely at some female mummies, they actually found very similar tattooed markings. The best example is said to be a woman from c. 2500 BCE Thebes with the name of Amunet, Priestess of Hathor. She had a number of clearly visible tattoos including lines on her thighs and arms and an elliptical design below her navel. These simple dotted and dashed designs were many times overlooked by archeologists. Then mummies from 1300 BCE turned up with Neith tattoos and 400 BCE women were found with Bes tattoos. Musicians and other female performers are frequently depicted with Bes amulet tattoos on their thighs. Seeing as how the designs were primitive at first and then later blossomed into tattoos that are still being used by women in Africa today, one can see too how henna body art likely also progressed slowly into the wonderful artform it is.

Even though henna body art can be certainly something to behold, beauty wise, it is not found prevalently in mythology as some suggest. Henna in general is hardly featured in any mythology, as is seen by the lack luster mentioning Culpeper provided in his book compared to Alkanet and other botanicals. One myth is that henna body art, as a form of cosmetic beauty (ornament), was discovered or first worn by Fatima. This is of course countered by not only literary sources but also the Koranic writings itself. Specifically the recounting of Mohammed's birth. Obviously women in the area of Mecca had to have been wearing henna stains on their hands as Mohammed was growing up in order for him to fall so deeply in love with it. Mohammed (c.600 AD) was even quoted as saying women who didn't wear henna on their hands were brazen. Fatima is also quoted as explaining women should adorn their hands with henna on a regular basis. So theories that Muslim women first started using henna on their hands during the Mogul period and specifically in India after Hindu women began to convert is quite silly. There is much, especially by way of poetry, suggesting women used henna on their hands prior to the founding of Islam. The same is true of patterned designs. Some have attempted to explain patterned henna first started in India among Hindu women and during the 16th century, spread all about the Old World. This isn't supported artistically however. They may point to the Ajanta caves but neglect to mention the art found there is clearly Buddhist. Miniature paintings dating all the way back to the 11th century have been found depicting Islamic women with discernible designs on their hands. Even more telling is paintings of Persian and Indian women depicted together in the 16th century whereas the Persian women have patterned designs. The Indian women simply have fully stained hands. One is actually hard pressed to find depiction's in India of real henna designs until after the Mogul takeover.

India suspiciously has a lack of a universal "Mehndi" creation myth like Muslim peoples have with Fatima. The main one revolves around Shiva's consort or wife Parvati adorning herself with henna to please him, thus Mehndi's linking to marital love, happiness and why it is used in the *wedding* setting. This is supposedly a "divya katha" or divine story. The problem however with such traditional myths is the way they have been constantly rewritten and transcribed. This may be a more classical or modern telling. Also English translations are frequently filled with errs. One would have to find the first instance henna was denoted in the myth. The actual traditional word used would also be very important, as many of the words pointed to as being indicative of henna really have a much older meaning of another plant. Likely the actual plant being mentioned in the ancient text. This is why henna appears to have hundreds of names in Sanskrit. Modern writers conveniently keep reinterpreting them to mean henna. Likewise some point out that henna body art is 1 of the 64 "arts" (Catuhsasti-kala) Indian women were suggested to learn in the Karma-Sutra. This however is in reality not as straight forward as it is presented and also not correct. In the lists of arts, one will find notations of a general "tattooing" and "coloring the teeth, garments, hair, nails and body". No mention is made specifically of henna being used in either case. Burton was very aware of henna body art (namely because he used it on himself as part of his disguise) and perhaps not knowing what the word "Madayantika" denoted, left it un-translated in the Karma-Sutra texts. It would be strange that if henna body art was either the tattooing or nail dye used, it would not be specifically mentioned as Madayantika. Furthermore, if it was so intrinsic to love in India, one would think it would be mentioned elsewhere in the book or some explicit directions given. The truth is that the tattooing notation is just what it implies, permanent designs being etched into the skin and not henna. Mehndi has become such a phenomena that many forget (or purposely neglect to mention) permanent tattooing was very popular (and still is in some tribes) among Indian women. In fact tattooing in general was considered in the *domain* of women and was performed by female artisans. The tattooing is believed to have spread from 2000 BCE Burma and was what Marco Polo (1275 AD) noted on his travels to East India. Assam Indian natives were seen with animal tattoos. I have seen examples of the Indian tattooing and the designs are what you would typically see for henna except using Indian ink injected under the skin. This is particularly done on the tops of the hands, wrists, arms, chest and neck. Just like what is traditional of ancient Visesakachhedya. Next we have the nail coloring and so on notation. This too is not indicative of henna body art. Instead it is a portion of the traditional "Solah Shringar" called "Dassanavasanangaraga". Otherwise it is comprised of the nail and teeth staining and so forth which used to be popular as part of a woman's Solah Shringar or 16 adornments. The area that, if henna was important in ancient India, would include henna in both the Solah Shringar and Kurma-Sutra would be "Angaraga". The Kurma-Sutra has a very interesting and inclusive chapter explaining ancient Angaraga (or cosmetics) but neglects henna, suggesting it was not popular at that time (c. 1st century AD). Other substances that were popular included sandalwood paste, alta, laksa and kumkum powder. For more on the Solah Shringar, read chapter 10. The basis for women using other substances prior to the introduction of henna probably intrinsically revolved around the traditional Tilaka. Today it is popularly known as the "Bindi" or red dot found on the forehead, between the eyes of women. The Tilaka has a number of purposes including denoting the 3rd eye and representing ones inner light or the sun in the Hindu religion. What many don't know is that the Tilaka was always a part of a woman's Solah Shringar and called Visesakachhedya which included placing various sizes of dots on a number of body areas. Not simply on the forehead. Also the Solah Shringar was not used only for weddings as it is now, in ancient times it was something a woman would attempt to perform daily. In addition to on the head, Tilaka's were to be placed on the chin, neck, chest and more interestingly, in the center of the palms. Such practices are frequently seen on Indian Tamil dancers today and religious statues. The red or yellow dot on the palm, which many claim as being indicative of henna in early murals, is likely a Tilaka and not Mehndi. This is also especially true of Buddhist depiction's, as they too strived to learn fine arts which would include those of the ancient Solah Shringar and they also wore bindi markings. Buddhism in general is a fine example of how henna body art was not present in very ancient India. Many think of Buddhism as being a strictly Chinese or Tibetan religion but in reality it started in India. The founder Siddhar the Gautama was born around 563 BCE and into the life of a privileged family in South Nepal (his father was a king). At the age of 35 he began teaching Buddhism and the creation of monkhoods. Mention of henna originally being used in ceremonies was not made. This is because Buddha likely died before it was brought to the area. This is heightened by the lack of notation by the Chinese who frequently made pilgrimages to India in order to obtain Buddhist scriptures. The claim that henna was integral to the early Buddhist comes not from the teachings of Buddha himself but instead pure speculation about what the colorant's are on the hands and feet of beings in the Ajanta caves (and other sites). Had henna really been integral to the Buddhist faith, it would have traveled much as it has in the Islamic religion. Areas where Buddhism has situated itself including Korea, China, etc., don't traditionally use henna to stain or adorn the hands and feet. The only linkage between henna and Buddhism comes from the traditional Buddhist Lent, where henna flowers were sometimes given to monks as offerings. This however was not an instruction by Buddha but instead subject to what commoner peoples had on hand in their particular area. Buddha apparently never instructed any of his monks or followers to color their hands. The only seeming mention of henna comes from the book "Banner of the Arahants" where a monk Ratthapala is quoted as poetically saying:

> "Behold a figure here pranced out
> With jewelry and earrings too
> A skeleton wrapped up in skin
> Made creditable in its clothes.
> Its feet adorned with henna dye
> And powder smeared upon its face
> It may beguile a fool
> But not a seeker of the Further Shore.

This however is an English translation of a translation, of a translation, etc. I wasn't able to see the word used in the original translation. Much of the Buddhist scriptures were destroyed in India so it is very hard to obtain pure originals. In any event, the manner in which Ratthapala speaks seems to cast a very bad light on the use of henna. Especially for those trying to be pious. I asked Buddhist monks about this and one explained to me, "What you are seeing (in Buddhist art from Nepal, Ceylon, Tibet, etc.) is iconography…like the halo on Western saints…in this case and I am working from memory here, the red hands signify power and compassion.". Color in the Buddhist religion is in fact very important and many times has significance unto itself which does not rely on reality.

FURTHER MISINTERPRETATIONS

Frequently Misinterpreted as Henna:
—Permanent tattooing.
—Any of the numerous other red, orange and yellow botanicals women used to create designs on their bodies including alta, laksa, tumeric, kumkum, myrtle, alkanet, indigo, madder, woad, etc.
—Reflexology depictions, especially in India, incorporate designs that are sometimes seen in Mehndi hand and foot designs.
—Deity markings, such as the eye of Horus for example is frequently symbolically placed on an Egyptian persons body. There are numerous other deity symbols which are frequently symbolically used in art but do not denote actual body markings.
—Visesakachhedya in India which is the ancient art of painting Bindi's on the forehead, chin, neck, chest and middle of the palms. This frequently done with kumkum and other red staining substances unrelated to henna.
—Veins in some ancient cultures were accentuated with a blue pigment. This was especially performed by women on the forehead, hands, breasts and arms to add beauty. Ancient Egyptian women were especially noted as doing such on the temples. If depicted in art, some may mistakenly perceive this as henna body art.

This is why it is very important when researching henna body art not to wear blinders or henna colored glasses. You must investigate everything as implied uses are not enough, especially in the midst of the many ways women have traditionally decorated their bodies with substances other than henna. An art history teacher and archaeologist in Athens explained to me about Greek frescos and ancient art, "We just have to keep in mind that all the colors that an artist was using were conventional. For example, they were using red if they wanted to depict a man, white for a woman or bluish for an area of the head. That is why, for me, the use of henna on the hands or feet consists an assumption. The artist may have wanted to depict something else.". He then went on that he wasn't able to find any depiction's of henna body art among Greek art but he did note reddened nails in some depiction's of women. Unfortunately, how do we know if the substance is cinnabar, alkanet or maybe henna? It is impossible to definitively ascertain. That is why I have used the written word as the basis for this book and art secondary to back up the literary sources. Not vice versa. For more detailed information on the history of henna as part of body art, please read my previous book "Mehndi: Rediscovering Henna Body Art" (Infinity PA: 1999).

HOW THE PASTE WAS CREATED

It seems like everyone's grandmother (in henna using locals) had a *secret* recipe for henna paste that outshines and outlasted everyone else's. Really, henna needs but two things to color very well and sink into the skin. That is acidity and water, or something to allow it to ferment. As long as you use dried henna however. Fresh henna simply does not work to produce Mehndi designs and is a myth perpetuated by those who never experimented with fresh leaves. I did a number of experiments using fresh henna recipes, including those presented by Dr. Mahendra Bhanawat and they all failed to work. This is because the drying of the leaves is what starts the chemical reaction for henna to produce good orange dye results. What is interesting however is that women didn't just use the leaves of henna for their paste. In Arabian areas women would dry and grind up the berries. In other areas such as Egypt women would use the flowers. Some would even use the roots or twigs. Loring (author of "A Confederate Soldier in Egypt") writes about how women applied henna in 19th century Egypt:

> "Before decorating the soles of the feet, already delicate among the refined, they are rubbed with a small instrument made of clay until they become still softer and smoother and therefore better fit to absorb the preparation. The dye is made of the flower of the henna tree, grown in Egypt and pulverized. When used it is diluted in water, afterwards rubbed on and covered for an hour. It then becomes of an orange color, which to the eye of the Egyptian is very beautiful."

In Egypt women were also said to mix the ground henna not with water but with fine oils and perfumes. Others, such as in India, were said to use lady fingers, alkanet, beetle-nut and all sorts of other naturally red dyeing substances to heighten the pastes outcome. For henna paste recipes that work, please read my previous book, "Mehndi: Rediscovering Henna Body Art" (Infinity, PA: 1999). For another method that does *not* work, read the words of Biblical critiquer Michaelis from 1709:

> "[The] custome of the Orientals of burning in their right hand all kinds of marks with the ashes of henna, which gives an indelible color and this is done even today."

Yet another reason why European men are many times not the best to rely upon for proper henna body art facts. It's quite likely Michaelis' words frightened a good many women from ever thinking of trying henna. This is also how outlandish myths start, metamorphoses begin and bits and pieces find their way into books of history. Similar to the myth that the Great Wall of China can be seen from the moon. Fortunately his words don't seem to have colored henna's history negatively.

THE MANNER HENNA WAS FREQUENTLY SEEN

> "Turks import small twigs of alkenna from Africa, with which they dye not only their hair but their hands and feet and nails upon them; and also other parts of the body, staining them a dark red."
>
> —*John Fryer, Persia 1681*

For many, the mention of henna body art invokes visions of highly decorated hands, as seen on lovely Indian brides to be. However, this is not a good example of how henna body art is worn daily by women or likely even how it was used 900 or more years ago. In reality, simply dipping the fingertips in henna &/or staining the palms was seen as being lovely enough to both women and men. There is little indication that having elaborate designs for everyday wear was a must. Women who had little work do to or who were very rich seem to be the ones to have created discernible designs. These women are likely the ones who started the trends of henna body art in the first place. Most likely, while living in a harem situation, they would attempt to outdo one another in order to get prominence and affection from their husband. Many hard working women who had to look after children, clean, cook, take care of animals and gardens likely had little time to create designs. Buck Whaley wrote in 1797 about Turkish courtesan women using henna in a patterned form:

> "They [Turkish women] usually paint their nails and eyebrows with a plant called kene [henna], which gives them a yellowish-red color. They sometimes paint the hands and feet, describing thereon flowers, etc."

Most writers such as William Loring noted the hands and feet of women were fully stained with henna, with no discernible designs. Many stated this made the women's hands and feet look dirty and dingy. Others said that their hands and fingers looked like something a vampire would have. Still others explained that the women would have been pretty had they not employed the very "disagreeable" looking henna. Edith Durham wrote at the turn of the century that Albanian women had "…their toe, as well as, fingernails were red with henna. All looked most unwholesome…". Orientalist writers were many times very cruel of the hennaed women. Adam Clarke (1810) seemed a bit different from the rest of the theological and Orientalist writers when he noted:

> "…more than one hundred drawings from life of Eastern ladies lie now before me. Their legs and feet; the soles of their feet and palms of their hands are colored beautifully red with henna."

This is likely because by his time, henna was found to be mentioned in the Bible and thus suitable for Christian women to use. Many people, including Clarke attempted to transpose the cosmetics used by women in Egypt and the Near East as being indicative of dress in Biblical times. Very infrequently will you find early notation of discernibly designed henna on women. Of course many will point to the miniature paintings and such but one has to remember that it was the rich that could afford these paintings. Not the general public. They represent only a small population of women using henna. Many women simply slathered henna on their hands, feet or other body areas while they were in the bath houses or for medicinal purposes at home. Fatima noted that women should use henna particularly on the hands to keep them soft, protected and feminine. That leads us back to the question of, when does henna stained hands and feet transcend medicinal use, into the world of body art? Especially when the results are one in the same. Another reason for a lack of designs (except perhaps on very special occasions, i.e. weddings, etc.) is that in the Islamic religion (as a whole) women are not allowed to publicly use ornaments that would draw male attention to themselves. In one story, a woman named Fatima (from the 10th century) enjoyed talking intellectually with an old sage. Even going so far as raising her veil in his company. One day she used henna on her hands (perhaps in an ornate, patterned manor) which caught the attention of the sage. He asked her about her hennaed hands which immediately ended their conversation. Fatima exclaimed that his attention to her hands denoted a desire for more than an intellectual friendship. Hennaed hands were obviously something only a husband and other women (and children) could admire. Many women were likely scared to bring unwanted attention to themselves and thus used only dip style henna or kept it confined to their fingertips &/or palms. In some Islamic cultures women can never expose hennaed feet in public and in very strict areas, hennaed hands are also not allowed around non-family members. However, it is well known Islamic women would henna their hands to discern themselves instantly (and even from a distance) from other religions of women. Many times non-hennaed women were attacked and assaulted by men, as they were considered of a minority. This would have certainly encourage Jews, Christians and other nomatic women to use henna as a measure of protection.

While Islamic and Hindu women are constantly pointed to as wearing henna, the medicinal and cosmetic use of *Lawsonia* spans all religions in the Old World. The only motive for not using henna was racial intolerance and a wanting to disassociate oneself from other women, perhaps looked down upon based on their religion. Christian women used henna just as Buddhist women did and other religions. The whole basis for henna's use stemmed from it's growing in close proximally and if the climate was hot enough to warrant its use. This is what led to Coptic, Catholic and many other Christian women to use henna throughout the AD period. Many Christians situated in Africa, the Near and Middle East and the Mediterranean likely understood the mention made of henna in the Song of Songs. The Greeks and Romans looked down on body marking (especially tattooing) which may have led to the abandonment of henna among some Christians. In addition, in those periods, henna was extremely expensive and in order to be pious, people were encouraged to abandon such luxuries and give the money they saved to the church. Henna was likely never a "holy" plant or substance to Christians, as it is to Islamic's, so as Christianity moved towards the colder areas of Europe, it was fully abandoned. It also appears that the Song of Songs, where henna is mentioned, became less and less understood or even discussed in the Christian society. Likely this was caused by language shifts as Christianity spread and Catholicism burning people for having a bible in their possession. Many people couldn't read, thus forcing them to rely on church sermons. This is seen in Bible notes by Geneva first published in 1599 where *all* of the passages containing henna are fully neglected to be printed. This would draw the conclusion it wasn't being mentioned at that period to worshipers. It wasn't until theological authors began going back and looking at the Latin Vulgate, etc. and taking trips to the area of Mecca and so forth, noting the actual plants growing in the area, that henna was tentatively found again. I say tentatively because there was still much confusion over what exactly "ko'pher" was. This all seemed to commence in the mid 17th century, particularly with Gill, well after the Spanish Catholic church banned henna's use amoung Christians. That was of course politically and racially motivated, not to mention an embarrassing mistake. After henna was found to be mentioned in the Bible (it is one of only three flowers in the entire book), it was associated with atonement and specifically Christ Jesus. Although some (such as the stigmata suffering nun Anne Catherine Emmerich) explined it was a flower under the guidence and protection of Mary, Jesus is more frequently associated with henna. Thus, it bacame a *wonderful* flower, sutible for all women to use, especially as a perfume. Unfortunately these revelations came too late for European and other women living in non-Islamic occupied areas. Westernization was already beginning to stamp out the, as some authors put it, "unwholesome" look of henna stained hands. Instead, gloves became all the rage and any type of apparent makeup was left to the harlots. Henna flowers and perfume were also extremely well know for being an aphrodisiac and addictive to women of Egypt, thus somewhat tarnishing it for self controlled women. By the time of Jamieson, Faussett and Brown's book (1871) and notations, henna was fading ever faster:

> Camphire or "cypress", the "henneh" is ment, whose odorous flowers grow in clusters, of a color white and yellow softly blended; its bark is dark, the foliage light green. Women deck their persons with them. The loveliness of Jesus Christ."

Their words showed a shifting to include the henna designs as being indicative of Christian worship as well. Likely because many Egyptian Coptic women used henna (as all other women did in the area) to stain their hands and feet. By this time however racial intolerance and a general disdain for non-white people by many prevented henna from being used in Europe as it has for thousands of years in Egypt and other areas. It also was so associated by Arabian and Islamic people in general, that once again, in order to disassociate themselves from them, many women abandoned henna. Instead of using henna for body art or even hair color, many women simply grew henna in their gardens, enjoying it in that manner alone. It however never transcended into Western

Christianity as being truly spiritual as the rose and lilly has, likely because it was correctly translated so late and was always perceived as being tied to Islam.

HOW HENNA DESIGNS APPEARED

> *"Alcanna being green, will suddenly infect the nails and other parts with a durable red."*
>
> —Sir Thomas Brown, 1650

The manner in which henna stains the skin seems very mysterious because it can take days for the true color to appear. This is seen on both the hair and skin. The reason for this is because henna's main colorant is Lawsone ($C_{10}H_6O_3$). Lawsone is actually very similar to the colorant obtained from the American native Black Walnut, which has the same molecular equation. Lawsone is also found in a few other plants including Jewel Weed which people have used to create body art. Many times in the exact same manner as Mehndi because they did not have henna on hand (no pun intended). Lawsone is almost colorless when it is housed inside the living (or just picked) henna leaf. Hence why when you brake open the leaf, no red coloring matter (liquid) is present. When the leaf begins to dry however, the Lawsone oxidizes and binds with other chemicals, which causes the darker coloration to start. If the henna is not properly shielded from air and sunlight, it will continue to oxidize (due to the movement of molecules) to the point of being unable to dye the skin or hair. When a paste of *dried* henna powder is created and applied to the skin or hair, the molecules of Lawsone adhere themselves into the protein (or onto the hair) and continue to darken further. Usually over 1 to 3 days. This is because they no longer have any leaf matter to protect them from the oxidation of the air. This is what allows the henna body art designs to gradually darken after the paste is removed. When the henna paste is left on, it acts as a shield over the Lawsone molecules which are trapped in the skin. When the paste is scraped off, the Lawsone is fully exposed to the air and continues to darken over a few days. Henna also has melanin stimulating qualities (similar to Tobacco juice) which plays a role in how the henna will react to each individuals skin tone and type. Thus, usually resulting in those with more melanin present to have darker henna designs. This is only seen when freshly ground dried henna is used however. Many people's dismal henna body art results comes from the use of stale henna and not leaving the paste on long enough. The major foe of Lawsone is water. If henna leaves are picked on a damp day, they will absorb water and produce a very bad henna powder. The reason for this is water can draw away the Lawsone molecules and prevent them from adhering to the protein or keratin of the skin or hair. This is also why so called fresh henna pastes (where fresh, un-dried henna leaves are used) don't work. The water added to the paste draws the molecules away from the skin and also disturbs the oxidizing ability of the Lawsone. I created a henna paste using fresh leaves and only received a faint yellow stain on my palms and feet which only lasted for but a 1 day. This is also likely why, traditionally, oil was applied over the designs to shield them from water. It has long been thought water causes henna designs to become ruined. While it may not fully wash away the designs, water does cause the designs to fade faster. The medicinal use of henna in a fresh form is frequently noted and the reason this was done, was to prevent unwanted body staining. Especially in the case of headaches and such. The dried henna paste has more of a glue like nature and thus was frequently used as a coolant by both men and women, irrespective of the resulting coloration.

THE WOMEN WHO PERFORMED HENNA BODY ART

With some Mehndi artisans having a backlist of clients that included music celeberties such as Michael Jackson and Madonna, movie stars such as Liv Tyler and Demi Moore, super models such as Naomi Campbell, it would appear to be a celebrated position. Especially since many artisans today are known for their expertise and personal style. This however is a modern, Western invention of the henna artisan. There is no evidence that women were held in any special esteem for being able to apply henna artiscally. No ancient henna artist priestises or set ceremony for its application have been found. No guilds or henna artist casts. This leads one to be reminded that if henna was thought as being medicinal, women likely applied it to themselves and thus didn't need a special applicator woman in ancient times. Likewise there isn't really any evidence that henna was used in a patterned manor prior to AD. That aside, the real women who emerge as the main ones engaged in applying henna were generally of a low class. Many were midwives who used henna medinally and cosmetically right before and after the birth. Other women were simple survants (and slaves) who did all sorts of chores including applying henna if need be. In India, many henna artisans were the wives of barbers, who are of a low class and frequently poor. Many of the people who worked with henna, including in the mills and in the fields were of a very low class and looked down upon, likely because the constant working with henna caused adverse health conditions. As is seen with many beauticians today. The only henna artisans that seemed to make a nice living was those in the royal court setting. Many times they were given expensive impliments to mix the henna paste in, which they either later sold or passed down to their daughters. Henna artisans appear to have always been exclusively women. They also exclusively appeared to work on other women. This was especially true in Islamic societies were men were (and still are) forbidden to use henna on the hands and feet as women do. This included on the palms. The busiest time for the henna artisan was likely at a "Henna Night" but since many cultures give the honor to the mother (or another family member), it is uncertain how long outside women have been used specifically for the application process. Once again, henna is a side job for some belly dancers and other forms of entertainment. It is unlikely women were called upon for their artistic gift with henna until well after the 10th century.

Chapter Ten

❃

Never Wear Henna to a Funeral...Unless Its Yours

The Use of Henna at Weddings and Other Joyous Occasions

"After the girls are betrothed, the ends of the fingers and nails are dyed red with a preparation from the Mendey or hinna shrub."
—Forbes, 1813

The wedding setting today is most certainly where you will see the most elaborate henna designs. Henna is usually applied during the traditional "Night of the Henna" which was heavily carried on by Muslim peoples. While today these henna designs are seen as diverting the "Evil Eye", being an aphrodisiac and having the ability to drive away evil spirits, the exact time in which henna became an integral part of a marriage ceremony isn't known. This is most likely due to the *fact* many wedding ceremonies in general, in ancient times, even in such places as Egypt, are not exactly known. The use of henna in marriage most likely came from simple origins. It could have come from women trying to cool themselves or a result of coloring their hair or nails with henna. One has to remember henna was extremely expensive in ancient times, especially if you did not live in an area that produced it. So weddings could have been the few times in a woman's life to use it at all. Henna was probably used as kohl and rouge was, to add beauty to the nails. Actual designs using *henna* most likely didn't start to be implemented until much later (such as in the past 1000 years). This can be compared to how modern brides wear one garter and go through a ceremony unto itself after the wedding, even though the original use for garters was practical (to keep the stockings up). Henna too has become a symbolic tradition that most likely had very practical uses at first. This is seen in a number of reconteurs from Turkey and India explaining that in the olden days they would simply hold a ball of henna in their hands, wrap the hand and remove the henna before the wedding. That would be the extent of the henna staining. The "Night of the Henna" as we see today also does not seem to correspond to how weddings were held in Biblical times among Jews and surrounding peoples. After having a procession around the town, the taking of the bride to the grooms home (tent) constituted the actual marriage. The *bride* would be in charge of cooking the food for the gathering that would take place afterwards. When everything was cooked, that is when she would pretty herself up, using kohl, jewelry and so forth. Fine oil / perfumes were extremely popular among richer brides. The most important period however was when the marriage was consummated and cloth stained with blood was given to the parents. As Deuteronomy 22:18 states, if a man ever tries to claim his wife was not a virgin when they married, the parents of the bride are to bring the evidence to the "older men" of the town. If they find the parents are correct, the groom received reprimand and was forced to pay the family money for disgracing his wife. Some have attempted to say the "Night of the Henna" traditions started in ancient Egypt but there is of course no proof of that. Others point to a mysterious 7th century BCE Assyrien text as stating girls used henna on their hands prior to marriage. I was unable to locate the above mentioned artifact but in order for this to be true, it needed to have been written in Aramaic. It would however make sense that where ever the traditions started, it did so in an affluent, modernized area with an extremely high regard for virginity and recognized marital union. This would somewhat discount Morocco and Berber women specifically as being the inventors of the Night of the Henna. In certain, remote Berber tribes (including the Tuareg) the women have a low regard for virginity / marriage and come and go as they please. No marriage ceremony or feast, some simply move in with the man of their choice and move on if they don't like the atmosphere. They are considered the most sexually liberated Islamic women. Most likely their behavior stems from traditional views of male and female unions. Similar to how certain tribes in South American rain forests conduct themselves. Only virgins would benefit from the female bonding and encouragement the Night of the Henna provides. For many it is the only marker of womanhood they will ever have. One must also remember that in ancient times a woman was *not* afforded a wedding at all. The actual first *ever* recorded auction was that of women for brides in Babylon. Women and girls would be sold off to men by the auctioneer and if not as a wife, as a slave. Such behavior continued throughout the Middle East and elsewhere right up to the Middle Ages. If you came from a poor family, had no dowry, it was to the auction house for you. One could forget having a wedding and everything pleasant that came with it. Then comes the *other* ways women were forced to be married. In India alone there is not one but 8 (Manam) recognized ancient types of marriage. India is always seen as the "henna-use-in-marriage" capital but in reality, fewer women then you think probably used henna for such purposes early on. The 3 out of 8 types of recognized marriage which certainly would have neglected the henna includes:

—Irakkatam where the bide is kidnapped or taken away from her family and forced to remain with the groom. Iratcacam is said to be the original form of such a marriage which was extremely popular in ancient India.
—Kantaruvam is what we would call a "common law" type of an arrangement. It would involve no wedding ceremony of any kind. The woman would simply move in with the man. This was said to be particularly popular with the Gandharvas.
—Paicacam is likely the worst type of recognized marriage which would result from an intoxicated or incapacitated woman being assaulted by the groom and thus shammed into staying with him.

Then one has to remember all the servant women, consorts and mistresses (otherwise kept women) all over the Old World who were never given wedding ceremonies. One certainly had to be somewhat privileged in the early days to have luxuries such as henna and fine oils included as part of the wedding preparations. The majority of the symbolism seen today regarding henna is a direct result of its incorporation into Islamic rituals and religion. Specifically its attribution to Mohammed and the Garden of Eden has led to henna being lucky, diverting the Evil Eye and having a blessing effect on the bride. So it is not surprising that the main ones who spread the Night of the Henna traditions were Muslims. From then on numerous other attributions were attached to henna in conjunctions to marriage, especially in India. There however is little evidence that the uses we see today are what were practiced in very ancient times. Henna being an integral, recognized part of a wedding party could be as little as 1000 or 900 years old. This is supported by how many traditional Jews do not practice a Night of the Henna ceremony. It is usually Jews that were either expelled from Spain such as the Sephardic, as henna body art was very popular there among all religions of women, or ones who live near Muslim women that have practiced a Night of the Henna. This is also true of the Yemenite Jews who live close to Muslim peoples and those of Morocco. It seems that many women, no matter what their religion, enjoyed using henna if they were in close proximity to Muslim women or in an area that grew henna profusely. Some have suggested that the Jews who moved to Europe, such as Germany and Russia forgot their henna traditions because they could no longer obtain or grow henna. Since Jews in Israel also don't have elaborate Night's of the Henna, this seems incorrect and unfounded. In addition, henna has been found to have been long used in the cold Caucasus mountains, Kosovo and by Latvian women so if henna body art was truly integral to a Jewish wedding, it's unlikely it would be simply forgotten. Many Jews probably had henna stained nails as the extent of their bridal adornment using *Lawsonia*. They also likely enjoyed using perfumed oil obtained from its flowers, which in itself is red.

Night of the Henna in Various Languages:
—Turky: Kina gecesi
—Afghanistan: Shao-i khinna
—Egypt: Leylet el-henna
—Arabia: Halat al henna
—Malaysia: Upacara berinai
—India: Mannziraath / Sanchit / Mehndiraat

The actual Night of the Henna can vary greatly depending on the local where it is held but the main components usually include music, dancing, food, folktales, female bonding, gifts and of course the henna. In some areas a great deal of henna paste is prepared for everyone to use. In other areas only the bride and her immediate family are allowed to participate. The henna designs done on the bride are usually the most elaborate in areas that have elaborate designs in general. Some cultures however forgo the designs and simply plaster the brides hands with henna. In some Christian gatherings only the pinkie finger is dipped in henna. Also popular in Coptic Christian wedding rituals is the use of candles to cause a blessing effect. The candles are placed in the thick henna paste (like a birth day cake), paraded around the home or room and then given to the bride to use on her hands a feet. The dance associated with this is called a "Zeffa". Perhaps this comes from the traditional use of torches / oil lamps by maidens during the wedding processions among ancient Jews. This could also be a root for why henna is important in wedding festivities, as the Evil Eye diverting and blessing effects of the candles carried over to the henna paste in which they were situated. In India the grooms initials may be hidden in the elaborate and dense henna designs and in some Islamic cultures prayers are written on the body with henna. Many Islamic based Night of the Henna's believe the designs have a blessing and evil spirit dispelling function. Hindu women feel Lakshmi hides in their designs and helps their marriage to be happy. Jewish women use henna because it is a natural colorant and some forms of Judaism do not allow women to wear cosmetics created with chemicals. In Spain the Moors introduced their henna traditions to Christian women who adopted using it in their weddings as well. This caused Emperor Carlos V in 1518 to publish a law stating Castilian women could not use henna on their hands or their feet and that they must get married in a Catholic church in a Christian manner. People complained about the laws but in an attempt to distance themselves from the Moors, complied. This however shows how quick women were to adopt henna as part of wedding preparations and how many tried to forget *who* brought the customs to the area, it order to keep practicing it. Turkish (and other) brides in the Caucasus seem to have been using henna since the Middle Ages according to the book "Soviet Anthropology and Archeology".

This theory is derived from very old epics of Aserbaijan called "Dede Korkut" which is also the name of the main character. It states in one of the stories concerning a wedding:

> "Her hair braided, wearing buttons of red, hands dyed with henna to the wrists, ornate gold rings on her fingers, the girl was married. The bride wore a scarlet veil."

Many women today sadly see henna as being something that is just done. Like new brides in the US wear white, hold a floral bouquet and make certain they don't let their groom see them for fear of having bad luck cast on their marriage. During Westernization and colonization, especially by the English, many affluent women abandoned henna, saying it was "old fashioned" and "too ethnic" to use at weddings. Fortunately tribal and poorer women kept the artform alive and it has once again returned to being an important part of many cultural weddings.

In Indian marriages an important aspect of preparing for the big day is to be certain to have all 16 items collectively called "Solah Shringar". Solah Shringar is related to "Catuhsasti-kala", or the 64 fine arts women are encouraged to learn about. Richer women that is. Henna *today* is a part of the 16 needed items. This however is an introduced beauty product as the Karma-Sutra and other books didn't originally mention it for such use. The Solah Shringar is typically pretty ambiguous, allowing for much leeway in the specific categories. Today they have however become very cut and dry. Typically they include (I am not presenting them in particular order) flowers for the hair, the bindi, jewelry, floral wreaths, floral necklaces, the sari, earrings, designs to be painted on the face, nose ring and so forth. The henna would be included in the section called "Angaraga" (also known as charanaraga) which is pretty generalized and includes all types of cosmetics, beauty treatments and perfumes. Of course originally henna was not included in Angaraga, as seen in the Karma-Sutra which is considered one of the best sources for traditional Angaraga information. The ambiguousness of the Angaraga category allowed for any type of perfume or makeup which was in vogue to slip in. This is what allowed henna to slip in much later. Originally women would try to incorporate the Solah Shringar into their daily lives, as an expression of fine art and their inner beauty. However today, most likely because of Westernization, it is frequently only manifested at a wedding setting. Afghanistan women are also said to observe something similar to the Solah Shringar called "Ha ft-rang" or the Seven Colors. Henna (khina) is included as one of the seven colors women should wear, specifically for decorating the feet and hands. Colors like blue (wasniah) from indigo are used on the eyes and red rouge (sorkhl) is used on the cheeks. Black of course comes from kohl and is used on the eyes and eyebrows.

For more on the use of henna in weddings, please read "Mehndi: Rediscovering Henna Body Art" (Infinity PA: 1999).

HENNA AT OTHER OCCASIONS

Henna in the form of body art has emerged as an important part of many happy gatherings and occasions. In India there are many times when henna is called for in addition to the numerous weddings held in December. Among Islamic women, the New Year is a popular time to grace the hands with festive color. Henna has become a medium of celebrating the joys of life, company and celebration. In some cultures girls are allowed to first use henna after their first or second menstruation. Moving them into a new world of womanhood. In other areas, especially in India, they will first appreciate henna in a symbolic nature during their wedding. For some this could be at the age of 2 or 5 years old. Women are certainly not the only ones to use henna as many think however. Islamic men would too color their nails for New Years in addition to their hair and beards with henna. Men of the Middle East were also said to use henna for vigor during warring times and in Morocco, captives were said to be hennaed before their demise. At certain times of the year, even cows in India get the chance to be decorated with orange henna spots during festivities. Sheep were symbolically adorned with henna before being slaughtered and cooked in some areas of the Middle East and brides would sometimes receive a wedding gift of a mehndied bovine. According to Dr. Bhanawat in his "Menhadi Rang Rachi" the main reason why women would use henna during festivals in Rajasthan was to insure their husbands would live longer then them. He recounts that women were frightened that if they forgot or neglected to use henna, ill will would befall their husbands and no children on their part. *Maybe*, but that seems to really take the fun out of the henna adornment.

Of course war is not really a happy time (unless you're on the winning side perhaps) but it has been made into a popular medium for men's henna history. Coloring the nails with red substances and making distinctive markings on the body appears to be a very old invention among soldiers. It however is also a very varied one that must be examined on an ethnic, religious and cultural level separately. Men in the Middle East have long been noted as having henna applied to their nails and at times other body areas by their wives for mostly magical reasons. Sir William Foster who edited Persian Travel notes of Fouruet (17th century) published:

> "They paint their nails and hands with alcanna, which besides the ornament it gives, cools the liver and in war makes them valiant."

Now this remark about at least 17th century Persian soldiers sounds plausible. However, in true Orientalist flair, the author can't stop and leave things be. He must flourish off to add depth to the subject at hand:

"Their nails are discolored with white and vermilion but why so, I can not tell, unless it be in imitation of Cyrus, who as an augmentation of honor commanded his great officers to tincture their nails and faces with vermilion, serving both to distinguish them from the vulgar and in fight appear more terrible."

For one thing vermilion is not one in the same with henna. Secondly, where is he getting this information? Herodotus (430 BCE) wrote that Ethiopian soldiers under Cyrus' grandson had their bodies covered, half white with chalk and half with *vermilion*, perhaps this is what he is confusing? What ever the case, over and over again especially regarding soldiers and supposed henna use the authors misinterpret vermilion and ambiguous *cosmetics* (war paint) as indication of henna. This is obviously what happened to the writings of Flavius Josephus who, according to one 19th century author, recounted in his "the Jewish Wars" that Jewish men wore henna in battle. When I read the book and war specifically, Josephus really said no such thing. In reality the dubious cosmetics was mentioned and not henna specifically. The passage wasn't even referring to soldiers. I have noted many Orientalist writers enjoyed embellishing to the point of being flatly wrong. They were many times good at recounting what they saw in their own present time but taking what they saw and then transposing it over ancient events seemed too irresistible. That is why it is very important to try and locate the origins of what they are explaining, stand it next to factual personal observations and see how it measures up. In Islamic law men are specifically forbidden from applying henna to their hands (palms) and feet in the manner women do. The only exception is if they have an urgent need to do so, such as for medical reasons I would suppose. This likely prevented many soldiers from using henna in such a manner in the Old World. In India, the warrior cast of men were said not to have used henna (specifically on the hands and such) because they thought it too feminine to do so. Usually boys and male babies are excluded from having their hands dipped in henna during their circumcision or purification ritual. Usually only the bridegroom is allowed to have henna applied before his wedding and not other family male members, except perhaps the best man.

HENNA AS A LAST RIGHT

Henna's use as a last right could have very well started in its floral oil or perfume form. Expensive oils were frequently used on the body of the deceased in many areas of the Old World, not just in ancient Egypt. The significance of such use may have spread to the henna dye which has been reportedly found on Egyptian mummies feet, hands and nails. It was not specifically mentioned as part of the embalming rituals (by Herodotus) but many historians believe henna was applied to some mummies (notably the royal) at the time of their death. Some also point to the Roman period papyri n. 5158 (housed at the Louve) and the Bulaq 3 (housed at the Cairo Museum) as noting henna's use on the nails during the beautification of mummies. While male (Pharaoh) mummies are usually talked about in conjunction to henna body staining, some Egyptian female mummies have also been reportedly recently found to have hennaed nails and fingertips. Other races of people having henna applied to them (and preserved in an Egyptian manner) have also been located including a Roman man with henna dyed hair. Another mummified person described as having hennaed hair (at least) is the Phoenician King Tabnit who was buried in an Egyptian fashion. Originally from 500 BCE Sidon, Tabnit's tomb and body was found in 1881; located in a cave named after Apollo in Mougharat Abloun which is present day Lebanon. Perhaps Tabnit confused some in thinking King Ashurbanipal had hennaed hair and such. In any case, he is housed in a museum located in Turkey and is described as being only skeletal remains today. Obviously who ever mummified him did not do a very good job &/or some say he was handled improperly and deteriorated over the years. On top of the skull is said to rest a few tuffs of preserved hair. It appears that the henna supposedly found in the hair was noted when the mummy was found, back in the 19th century. Very little seems to have been written about the mummy itself and it is doubtful microscopic research has been done on the fragile hair yet. Many times the Victorians would look at *any* examples of reddened hair as being representative of henna dye. Many didn't consider the ancients as actually having natural, non-black hair. Since hair can turn very red due to oxidation, it is uncertain exactly what has caused the red color noted in King Tabnit's hair. Since Pliny explained Sidon was producing very good henna perfumes in at least his time, it could be very well that the curly hair (if it is truly his) on Tabnit's head is henna dyed. A case by case basis is always needed however and it is of course impossible to use pictures to determine anything dye related. Yet another hennaed man of high rank was in the discovery of a Fatimid burial ground in Cairo Egypt from the Shiite dynasty. In a large mausoleum, a man was wrapped in a number of layers of "tiraz" (colored fabric reserved for the Caliphs) and found naturally mummified. His preserved beard was said to have been dyed with henna. This led the French archeologists to assume he was of importance, as wearing henna on the beard is still a status symbol for men in Egypt today.

Henna is usually perceived as a joyful botanical, for use in lighthearted and happy events. This being so many women will forgo any type of henna when a loved one or spouse has died or when they become a widow. Islamic morning rules specifically forbids the use of henna among the living family members. On the dead, its first main functions were most likely an attempt to stifle the odor of the rotting corpse as opposed to apparent henna marking. This is seen in Africa where a morning powder is made with dried and crushed henna flowers (along with other aromatic plants) and sprinkled over the body. Later, the henna stains became a way of adorning the deceased. This was particularly done on women, especially among nomads, and confined to the hands, feet and nails. Women on their death beds were also many times adorned with henna and thus no further application was needed after their death. This would have also allowed the color to develop better as heat in regards to living tissue allows the henna to stain much brighter. Men in India would sometimes have henna applied to their forehead after death. In some Islamic cultures it was thought the henna would make one more pleasing to the angels and thus assure them getting into heaven. Many of the female angels in Islamic culture were thought to wear henna on their hands as well. For more on henna's use in death, be sure to read chapter 4 and 13 of this book.

Chapter Eleven

❋

Henna in Art

From Belly Dance to Illuminated Book Pages

> "Her hand is tinted gold with henna,
> She gave me a cup of wine like gold water,
> And I said the mood rise, the sun rise."
> —Hefny-bey-Nassifm, In a Yellow Frame

Due to henna's exotic nature, many authors included it in their poems and works of fiction. Here I have compiled a few poetic verses which make mention of henna for you to enjoy. This is over and above the many poems already strewn about this book as a whole. Note that henna (or its synonyms) is present. Poems simply speaking of reddened nails, hands &/or feet are not included because an assumption must be made. Being that many red cosmetics were used by the ancients, one runs a very high risk of being wrong when associating them with henna.

"Eyes with kohl darkened, hands with henna crimson
dyed; Through these beauties vain and wanton
like to thee was ne'er a bride."
—Fuzuli, 16th century

"Henna makes pure and good the hand's acts and
eases the path followed by the feet."
—Arabian Nights, Translated by Burton

"I'm sure she is the daughter of the Emperor.
Her nails are stained with henna.
They are like the petals of a rose."
—Oscar Wilde, La Sainte Courtisane, 19th century

"What do you weigh, O ye vendors?
Saffron and lentil and rice.
What do you grind, O ye maidens?
Sandalwood, henna and spice."
—Sarojni Naidu, The Bazaars of Hyderabad, 19th century

"Her nails were touched with henna;
but again the power of art was turned to nothing,
for they could not look more rosy than before.
The henna should be deeply dyed to make,
the skin relieved appear more fairly fair."
—George Byron, Don Juan, 19th century

"My hand is henned, my head veiled,
I go a little, I go well.

> Wake up Hadji Bey, wake up,
> I have come here a bride and I will go back a girl."
> —Turkish Wailing, Hadji Bey

> "She tinged the inside of her hands with Lawsonia, spread vermilion upon her cheeks and antimony along the edge of her eyelids and lengthened her eyebrows with a mixture of gum, musk, ebony and crushed legs of flies."
> —Gustave Flaubert, Stammbo 1862

> "Kabeer, I have ground myself into henna powder. O my husband Lord, you took no notice of me; You never applied me to your feet."
> —Sri Guru Granth Sahib, Sikh Holy Book

> "Amid the eddy of these dream fragments,
> Amid the smell of henna and the twanging of the guita,
> Amid the waves of air…."
> —Tagore, 20th century

Henna also made its way into a number of traditional songs. Many of which are sung by women around the Night of the Henna gathering. In some areas of India, it was thought that unless henna songs were sung, the henna would come out very poorly on the skin.

> "Daughter let your henna be happy,
> let your talk be sweet.
> Call her mother to come here,
> let her see her daughter become a bride."
> —Turkish "Night of the Henna" song verse

"A Kokila called from a henna spray,
Lira! Lira! Lira! Liree!
Hasten maidens, hasten away,
To gather the leaves of the henna tree.
Send your pitchers afloat on the tide,
Gather the leaves before the dawn is old.
Bring them in mortars of amber and gold,
The fresh green leaves of the henna tree.

A Kokila called from a henna spray,
Lira! Lira! Lira! Liree!
Hasten maidens, hasten away,
To gather the leaves of the henna tree.
The tilka's red for the brow of the bride,
And betelnut's red for the lips that are sweet.
But for lily like fingers and feet,
The red, red of the henna tree."
—Sarojni Naidu, In Praise of Henna Indian Folk Song

"Sing for the brightness of the white flower,
Oh, the father of the flowers.
Oh, the father of henna,
I will not leave you alone."
—Jewish "Night of the Henna" song verse

Dancing was not surprisingly another realm for henna to implant itself. This is especially seen about the Night of the Henna, where at times women are paid to dance as entertainment. One of these traditional dancers, which strangely enough is neglected to be done in India anymore, but still performed in Malaysia is "tari inai". Here dancers (traditionally women or girls) assemble and dance to instrumental music. The dance became so stylized and well known in Malaysia that it began to be performed for non-wedding related events including festivals, royal meetings and dignitary visits. The "inai" henna dance was believed to have been once reserved for rich royal weddings but after Western influences hit, became driven to the realm of simple folkdance. Thus it has become merged with real folkdances of the area. The dance that is thought to be pure compared to others is called the "mak yong inai". Another form of the henna dance is called the "Perlis inai" or "Terinai" dance and involves young women dancing with candles. It also includes one male dancer. Such uses of candles during the Night of the Henna is seen in other countries as well including Egypt. Called the "Zeffa", it was performed by either a paid dancer or family members and would include special wedding night songs and candles mounted in a bowl of henna paste. About the bride the dancers would pivot and this was said to bless the bride and divert the all too menacing Evil Eye. So comes in Oriental dancers, better known to most as belly dancers. Many have noted them to be frequently depicted with henna stained hands and bodies. This is not surprising because drawing attention to the hand and feet movements was important, especially from a distance. Mention of priestess type dancers using henna however have not been made. A great deal has been written by many about temple dancers but henna was not, suggesting it was not in the domain of such women. It wasn't something intrinsic to their dances. However, where available henna would be used like other cosmetics such as kohl to attract onlookers. Another reason many musicians and dancers appeared to use henna more than other people is because they thought it would make their hands and feet tougher. Thus better apt at holding up for hours and hours of continuos use. Stringed instrument players (male and female) frequently dipped their fingertips and nails in henna, to make them stronger and play better. It's very likely dancers of all types used henna for the same reasons throughout the ages. Edward William Lane wrote in the 19th century about henna use among Egyptian street dancers (Ghawazee's):

> "They also ware various ornaments, their eyes are bordered with kohl and the tips of their fingers, palms and their toes and feet are usually stained with the red dye of the henna, according to the middle and higher classes of Egyptian women."

HENNA AND ART

Instead of attempting to interpret ambiguous red markings on artifacts as henna, one is much more apt at finding the truth by studying preserved fabric and paint remnants. Such as frescos. Henna has very specific indicators when looked at under a microscope and its traditional use as a dye and its tenacity would allow for it to be observed on even very ancient artifacts. As has been seen with mummy hair samples. Henna as a clothing dye would likely have been reserved for poorer peoples. Such as perhaps the Israelites. A misconception many have is that henna dyes very red or orange as is seen on the body or hair. Without a mordant or additives, it's dyeing capabilities are quite poor. I dyed cotton and wool with natural henna and only was able to get a dingy yellowish color. Heat had little effect on the color outcome and applying the paste to the samples also failed to make a difference. When I washed the samples, much of the color went down the drain. Therefore much work is needed to be done in order to successfully use henna as a textile dye. Specifically by way of a mordent. Many of the mordents needed were likely not used by the ancients, such as tin which turns henna deep reddish brown. Typically, all dyeing was done using the dye bath method. Painting the henna on as one would for body art was likely not performed by the ancients. With good reason too. The green chloroform in the crushed leaves created an awful stain on some of the samples and I had to really scrub to get the muck off the fibers. Instead, dried henna leaves would be cooked in water to bring out their color and then either the cloth or yarn would be pretreated with a mordent or an additive was mixed into the henna brew. It seems also that the variety of henna used was important. In Turkey at lest, henna used for yarn dyeing is called "iplik kinasi" and in India the Rajani henna is said to be better at dyeing things. Among textile manufactures, *L. inermis* was thought to a give a deep orange or red color. *Lawsonia inermis* in general is thought of as being true "red henna". *L. alba* on the other hand was used for a yellow or orange color. *L. spinosa* was used simply for orange and so was *L. mimata*, which was perhaps a Japanese species of henna. To make a very black or gray stain, I used ferrous sulphate. That worked very well to stain the cotton and wool. Indigo was frequently combined with henna to create a black fabric dye that didn't need a mordent. Traditionally a deep brown was made for wool that had been mordered with cream of tarter or potassium dichromate. For a nice orange color, either for silk or wool, aluminum sulphate was needed. Other additives included bismuth, cobalt, salts of aluminum, iron, lime and so forth. Henna was popularly used for the creation of rugs and other textiles that would take a lot of wear and tear. Many mummy shrouds are noted as having henna staining present. Any dye used in ancient textiles should be carefully inspected for the presence of henna as it can be very telling as to if people in a specific area were in fact using henna in a number of capacities. This can be done via sampling and microscopic analysis. Many textiles in the Middle East literally have henna designs printed on them, especially as a border. They usually appear like 4 petaled, white, star looking flowers and is traditionally called a "Gul-i-henna" or "Gulhina" pattern. Green henna leaves will usually accompany them in a vine formation. Henna flowers are frequently found incorporated in floral rugs in general.

Microscopic analysis can also be done on paint chips to determine if henna is present. Likely not well known, henna can and has been used as a medium for more permanent art (in comparison to body art). This includes staining rawhide on drums, lamps and other artifacts in addition to creating paints. Typically henna has been used to create orange colored paints used specially for tempera, fresco, encaustic and even oil paints. It also can be used to make a brown, semi-translucent medium. It may have been mixed with the roots of borranginea or alkanet to create varying shades as well. Paint sampling has become a very advanced science and it would certainly tell a lot about ancient use of henna in any given local and time period. Tempera type paints were used since very ancient times and adhere to many types of items including wood very well. It is usually somewhat translucent, as it is made with egg and needs to be painted over white surfaces. Fresco on the other hand likely produced very vibrant orange results as the henna would react with its base of lime. This should be closely inspected in ancient Minoan, Greek and Roman art for the possible presence of henna which would help in dating *Lawsonia spp.* use among the peoples. Encaustic paints are created using wax or resin as the binder and is thus very durable. Such a medium is thought to have been used in the ancient Egyptian murals found in tombs. Henna does not impart orange or red results in wax and resin however (unless the roots were used). Instead the resulting color would be green or a brownish green. That would have also likely been the case for the oil paints as well. I was able to create a very black ink using the Middle Age gall ink recipe of Theophilus; of course substituting the henna leaf powder for the Oak Apple powder. It worked very well and was very stable on the heavy weight white paper I used. It was also very fast to dry and I didn't need to add any gum to thicken it. Perhaphs there are a number of manuscripts from henna growing areas that really use such a *Lawsonia spp.* based ink. Henna however is not the bases for "Persian Red" as I have seen some note. Instead, Persian Red was originally extracted from the madder plant before it became commercially produced from chemicals. Any type of an artifact can be tested for the presence of henna and should be, to give crucial clues as to the real use of henna in various ancient cultures.

Likewise, animal remains can also be tested for the presence of henna. Many people, including Porta noted that the mains and tails of horses (particularly that of Persian and Arabian breeds) were stained with henna. Numerous items were created by the ancients using horse and animal hair. Since henna has a great affinity for protein, it fashions itself to horses hair particularly well. Stained horses hair items which look red or orange should certainly be tested for henna deposits. All manner of furs can be treated with henna as well. One 19th century author noted her white cat had been dyed with henna even, while on assignment. Sheep, cows and goats have also been frequently mentioned as being hennaed. However, this can come from some religions not wishing to brand the animal and thus, instead use the henna to mark their flocks. This was especially done on sheep because their thick wool would mask branding marks anyway. What this all shows is that archeologists can test many items for the presence of henna dye. It would certainly help to cast light on henna's use among ancient peoples of the Old World. People need to relieve themselves from an exclusive fixation with henna body art and look for other forms of artifact evidence that may help to answer the questions as to how henna spread from country to country. Right now the best source for such information comes from the linguistics and etymology of henna's name. This can of course change if studies are funded to do scientific testing on artifacts. The care that the archeologist or anthropologist takes with the artifact is also extremely important. Today, I am very pleased to see that even jars taken from ship wreaks are carefully kept in order to preserve what ever may still be inside. 10 years ago it would be normal to see people simply pouring out the contents of the jars right into the dirt of the dig and moving on to the cleaning process of the object. Henna in powder form and especially when old and stale looks very much like dirt. The evaporated oil may leaves only a residue. The dried flowers may have weathered away into brown, crumbling masses. With added care, it is likely true henna artifacts will turn up ever more all over the Old World. In doing analysis of artifacts, laboratories should look for the presence of $C_{10}H_6O_3$ Lawsone or Hennotanic acid. This can be extracted from paint, hair or fibers using various acids and carbon tetrachloride. A fast test may be to place paint chips or other types of stained samples in a solution of water and ferric sulphate. If the sample turns ink black, it may be detecting the presence of henna. In the circumstance of henna, it is really impossible to definitively state its presence without lab testing. There are simply too many oxidative causes of discoloration as well as other botanicals that could have been used instead.

Chapter Twelve

❀

Camphire or Camphor?
The Confusion Over Lawsonia inermy

> *"Like Perseus saves her, when she stands*
> *'Exsposed to the Laviathans.*
> *'So did bright lamps once live in urns,*
> *'So camphire in water burns,*
> *'So Aetna's flames do ne'er go out..."*
> —Ann Killigrew, To The Queen 1685

Many whom have read the King James Version of the bible is likely to know the given name for henna found there, as being camphire. At the finding of this supposed synonym for henna, I quickly began to research old texts which contained reference or recipes containing camphire. Finding such writings was intriguingly easy and plentiful. One thing I have learned from researching henna is it does not give up its secrets easily. What I later discovered was instead of an ancient name for henna, a large mistake had been made by rewriters of the Bible. Some have insisted camphire is a biblical name for henna and/or henna's Latin name. Some also said it was derived from the Hebrew word "ko'pher". All these theories are incorrect, as we will soon see. The only two passages in the bible which refers to henna is in Solomon's Song. The modern day translations can be found in this book (Chapters, 2 & 3), but the KJV, also known as the authorized version, reads as follows:

> "My beloved is unto me as a cluster of camphire in the vineyards of Engedi."

> and

> "Thy plants are an orchard of pomegranates, with pleasant fruits; camphire with spikenard, spikenard with myrrh."

Prior to the KJV bible, the Latin Vulgate by Jerome in the 4th century AD, was one of the first complete translations to be mass distributed. Completely written in Latin, Jerome, whom was born in Turkey, used the actual Hebrew scriptures to create his new bible translations. He rendered Solomon's songs as such:

> "Botrus cypri dilectus meus mihi in vineis Engedi."

> and

> "Emissiones tuae paradisus malorum punicorum cum pomorum fructibus cypri cum nardo."

Note how *camphire* is absent. The "cypri" is what stands for henna in at least Jerome's 386 AD Latin texts. Since he grew of up Turkey, he likely had a better understanding of what henna was compared to European scribes. Camphire does not make a presence until the bible is translated again, this time not into Latin, but English for the KJV. While the Hebrew name for henna "Kyphor" (ko'pher), which originally appeared in King Solomon's song sounds somewhat like camphire, it sounds and looks even closer to another name very well known to English writers, camphor. It's extremely likely that the translators mistook "Kyphor" for camphor (*Cinnamomum camphora*), as nowhere else *except* the KJV bible, does camphire denote henna. Cypri, the original Latin name for henna, was also mistranslated as meaning the Cypress tree in addition to the island of Cyprus. Translative mistakes do happen. The key is catching them before history becomes further denatured. In numerous, old English books you will find the word camphire next to a description of a white, hard, cool substance. Otherwise camphor. Jonathan Warren, was kind enough to send me a compilation of what he found about the truth about camphire. I originally contacted Mr. Warren after reading a very interesting article he had written, which included camphire and a very camphor sounding description. He replied saying he wasn't aware that camphire was ever used to denote anything but camphor or that the words were not one in the same. He provided the following interesting insight:

> "Indeed, the various historical dictionaries that have been incorporated into the Early Modern English Dictionary database (the project I was a part of and which inspired the article to which you refer) include numerous entries which conflate the two terms and none which distinguishes them as names for discrete materials. Thomas Thomas' Latin-English dictionary of 1587, for example, defines the Latin "camphor - a, -ae" as "The gumme called Camphire." John Florio's Italian-English dictionary of 1598, defines the Italian "Canfora" as, "a drug called Camphire. Also a kinde of tree or wood thereof." Similarly, John Minsheu's Spanish-English dictionary (1599)—one of the dictionaries with which I worked closely—translates the Spanish "Canfora" as "camphire."

Randle Cotgrave's 1611 French-English dictionary offers the following: "Camphre: [m.] [The gumme tearmed, Camphire.] [Camphre artificiel.] [Artificiall Camphire, is such, as hath beene refined, and whitened in the Sunne, or by fire.] [Camphre en rose.] [Naturall Camphire, is such, as hath not beene touched by fire.] [L'odeur de camphre chastre l'homme:] [Pro.] (Such power hath that Simple, diuers wayes, to make a man chast.)." Here is the entry from John Bullokar's 1616 English hard-word dictionary (the one I mentioned in the article in this regard and the one you asked about): "Camphire. A kinde of Gumme, as Auicen wri teth. But Platearius affirmeth it to be the iuice of an herbe. It is white of colour, and cold and dry in operation." I followed the tendency of Bullokar's contemporaries (and later lexicographers) by modernizing my reference to Bullokar's entry with the term "camphor." Interestingly, perhaps, by 1617 (in Robert Cawdrey's dictionary) and 1623 (in Henry Cockeram's English hard-word dictionary), "camphire" was not defined with reference to camphor, but merely as a kind of "gumme" (Cawdrey) and "a gum very cold and drie" (Cockeram). I cannot say whether this failure to mention camphor represents a shift in lexical understanding. However, as you perhaps know, the current Oxford English Dictionary's entry for "camphor" (there is no similar entry for "camphire") notes in its etymological summary, "Various forms of the word occur in 16th c. Eng., but the typical form down to c. 1800 was camphire; the mod. camphor is conformed to the Latin" (OED Compact Ed. i.324). In this way, the authoritative source on English etymological history solidifies the merging of the two terms. Another complication: you mention that camphire is not gummy while camphor is. Many of the dictionaries note that "camphire" is a white gum. It is conceivable, of course, that variations in spelling merely confuse the fact that all these dictionary writers were referring to what we call "camphor" and did not know of the henna substance called "camphire." It would hardly be the first time that orthographical variation led to such confusion."

This is brought out even more by herbalists such as Culpeper and Mr. French, whom describe in detail that camphire is camphor. It is also seen in 18th century chemical terms where next to camphire it stated "see camphor", as if they were one in the same while Lawsonia inermis is described as a plant used as a cosmetic, "The dried powdered leaves were used as a dye or, with suitable medium, a cosmetic". Camphire/camphor/camphora is described in this same book of 18th century chemical names as, "An aromatic extract from the sap of certain trees found in Brazil and the Far East", such as China. If camphire was henna, a connection would have been made. Camphire followed by "see camphor" is quite a normal occurrence in books, so it can be said that most people thought of camphire as another name for camphor and not henna which was called "al chena", "alcharma" and "al henna" at those periods, as seen by 17th century works by Culpeper, Porta and others. Mr. John French wrote about camphire in his 1651 book "Art of Distillation" and brings some of the most compelling evidence that with exception to the KJV bible, camphire always denoted camphor. This in seen in recipes including "Oyle of Camphire":

> "Take of camphire sliced thin as much as you please, put it into a double quantity of Aqua fortis or spirit of wine, let the glasse having a narrow neck be set by the fire, or on sand or ashes the space of five or six hours, shaking the glasse every

half hour, and the camphire will all be dissolved and swim on the Aqua fortis or spirit of wine like an oyle.

Note that if you separate it, it will all be hard again perfecntly, but not otherwise."

French speaks classic camphor, a hard white gum that can be boiled to make an oil or left hard to slowly evaporate over time. In each of his recipes, there are three, he spells camphor, camphire. We can also see how "camphire" begins to appear in English writing extremely frequently after the KJV bible of 1611. The first mention of true camphor is believed to have been by Aetius, a Greek medical writer of about 540 AD who calls it "Caphura". Note that his works were in Latin. Later, in 940 Masudi speaks of a country called Kansur being famous for its camphor. Then, all of the sudden, "camphire" starts to be used to denote camphor including writings from ship logs and herbalists like Culpeper. At the same time, camphor was being used as well and even other misspellings, such as "camphyre". According to the 1913 Websters Dictionary, camphire was an old *spelling* of camphor. In fact, a very old ship log shows camphor being spelled camphire prior to the KJV Bible. This is seen by the numerous allusions in literature including Ivanhoe and especially poetry which notes camphire's pungent scent. Camphor (camphire), was the original moth ball and was frequently used as smelling salts to wake those whom had fainted. Some (who likely never really smelled henna flowers) in the 18th century even said henna flowers smell like camphor, likely making the confusion even worse. John Gill in the 17th century however did write, "nor can it [ko'pher] be supposed that what we call "camphire" should be ment, which grows not in clusters, and was unknown to the ancients." Thus providing further indication "camphire" always was a synonym for "camphor". It's history is quite colorful like henna's but also has a darker side, as in large amounts it can be poisonous to the human body. In a Medieval Islamic manuscript, it was warned that people cutting the trunk of the tree should be careful not to allow the camphor to spray in their face. People were said to have been killed in that manor. Signs of such poisoning include a flushed face, dilated pupils, sore throat and a wreaking of the camphor odor about the body and breath. Studies on frogs showed increasing paralysis and convolutions from poisoning. Traditional treatments, which probably didn't do much of anything included extremely strong coffee, keeping the head cool and the body warm, bromides, etc. Of course, today a trip to the hospital or call to the poison control center is in order. It is very easy to become overcome by spirits of camphor. True camphor ($C_{10}H_{16}O$) is obtained from the *Cinnamomum camphora*, otherwise the camphor tree, and is a white, crystalline compound which is quite volatile. The people of Bengal would burn powdered camphor for religious ceremonies and the oil too was burned for fumigation purposes. Native to Asia such as China, Taiwan and Japan, it was medicinally used for a number of conditions related to colds and illness. Culpeper, in the 17th century wrote:

> "Camphire eases pains of the head coming from heat, takes away inflammations and cools any place which it is applied."

Since Culpeper also speaks of henna in his book, yet makes no association with camphire, shows camphire and henna were not one in the same to the minds of 17th century or onward peoples. Culpeper frequently researched the ancient Latin texts of other naturalists and herbalists, so if camphire and henna were thought as one, he most certainly would have stated such. John Baptist Porta, in his book "Natural Magick", also speaks of henna and camphire separately. He writes the following about camphire/camphor when he speaks about anise seed oil:

> " It congeals in water like camphire."

Porta, always experimenting, also stated that camphire was placed in the mouth of a roasting bird (chicken, duck, etc.) and when served, would cast flames out of its mouth. This was seen as an appropriate and exciting way to serve food to guests. Camphor is very volatile and like Ann Killigrew's poem suggested, can burn even when suspended in water. What this all means however is that poems such as "Upon Julia's Unlacing Hersealf" by Robert Herrick and that of Ann Killgrew's don't really mention henna but instead camphor. Perhaps the only exception would be Christian religious writings of early times. Therefor camphor should not really be noted as a well known synonym for henna but instead, a mistake of translators of the Bible. The notable 17th century English writer, Sir Thomas Browne, sums up the situation nicely:

> "…why it should be rendered camphire your judgment cannot but doubt, who knows that our camphire [camphor] was unknown to the ancients and no ingredients into any composition of great antiquity. That learned men long conceived it a bituminous and fossile body and our latest experience discovereth it to be the resionous substance of a tree in Borneo and China and that the camphire that we use is a neat preparation of the same."

HENNA'S COUSIN ALKANET

> *"The roots of these are vsed to color sirrups, waters, gellies and Fsuch like confections as Turnsole is."*
>
> —John Gerard, 1597

Yet another source of confusion comes from a little known botanical in the west, called alkanet. What is even more confounding is the fact, "alkanet" is derived from the name "henna". Beginning around the 14th century, people started calling *Alkanna tictoria, Anchusa officinalis* and even real henna "alkanet". Related to the borage and a whole host of other plants, the name alkanet is Middle English which was derived from the Old Spanish word for henna, "alcaneta". Alcaneta was a diminutive of alcana, meaning henna plant and itself derived from the Medieval Latin "alchanna" and Arabic's "al-hinnA", both meaning henna. Previous to alkanet, as noted by Culpeper, it was called "Enchusa" and "Anchusa" in Latin, which is derived from the Greek word, "Anchousa", meaning "paint". If the name of this plant is not confusing enough, its use throughout history is quite similar to henna's with one exception. It was edible.

Alkanet, which is also called "false-henna", is a plant which produces bright red dye from its carrot like roots. Unlike henna, it is very apt to grow in colder areas and is believed to have originated either in Asia or Europe, such as Russia. Even though this is the case, the Egyptians wrote about Alkanet in ancient texts, calling it "nesti" and so did Pliny and others of his time, noting its use to color perfumes and as a moder for cloth. Likewise species of *Anchusa* are found and considered native to the Mediterranean, Turkey and Israel. In many parts of India, Alkanet which is called "Ratan Jot" in Hindi, is thought of as being equal to henna and is frequently used along with henna or *alone* to produce Mehndi designs on the feet and hands. A number of traditional "fresh" henna recipes used for body adornment (staining) from India calls for alkanet. The reason for this is fresh henna leaves stain yellow and tan. Without the addition of intensely red ingredients, henna pastes using fresh leaves do not produce classic orange-red results. At certain times of the year, dried henna may not have been at hand, prompting the use of fresh leaves. Such as with rainy seasons. Alkanet is also frequently used in curies and other Indian dishes to give an extremely rich red color. Unfortunately some western writers have misinterpreted Alkanet's use in food for true henna, which can be dangerous as henna should not be used in food. If you ever note a recipe which states a pinch of henna needs to be added, interpret that as being alkanet. Alkanet really shouldn't be used either in food anymore as cancer causing properties have been found and it has been banned by the US FDA as a colorant.

Like camphire, many of the ancient writers speak of alkanet yet make no connection to it and henna, except for occasionally listing alkanet and orchanet as a synonym for true henna. One to have the most to say about alkanet is the 17th century English herbalist Culpeper. As we remember, Culpeper used ancient texts frequently to form his biography of a plant, including that of Pliny Elder and others. Many times this would include reference to mythology, which is lacking from his description of henna. Most likely because he couldn't find any. He writes the following about Alkanet:

> "It [alkanet] is an herb under the dominion of Venus and indeed is one of her darlings, though somewhat hard to come by. It helps old ulcers [wounds], hot inflammations, burnings by common fire and St. Anthony's fire by antipathy to Mars for these uses your best way is to make it into an ointment. Also, if you make a vinegar of it, as you make a

vinegar of roses, it helps the morphew and leprosy. If you apply the herb to the privates, it draws forth the dead child. It helps yellow jaundice, spleen and gravel in the kidneys. Dioscorides saith it helps such as are bitten by venmouse beast. It stays in the flux of the belly, kills worms, helps the fits of the mother. Its decoction made in wine and drank strengthens the back and eases the pains thereof. It helps bruises and falls and is as gallent a rememdy to drive out the small pox and measels as any is; an ointment made of it is excellent for the green wounds, pricks and thrusts."

As we can see, the uses are much the same as henna, medicinally speaking at least and that is because of the Law of Signatures. Culpeper speaks of alkanet bringing forth the "dead child" and this most likely is taken from the writings of Dioscorides whom states alkanet was used by women of his time as an abortion inducer via a suppository. Alkanet is extremely unsafe for pregnant women and should not be used in any form at that time of life. Back in Culpeper's time however, it was given to help speed up the birthing process. A 1552 book called "Le tiers livre des faicts et dicts Heroiques du bon Pantagruel" by French physician Francois Rabalais makes mention of alkanet and Dioscorides as saying the following:

> "Anchousa [alkanet], which some call Calyx, some Onoclea hath leaves like to the sharp-leaved lettuce, rough, sharp and dark. The root, ye thickness of a finger of ye color almost of blood in ye summer becoming severs, dyeing of the hands."

The dyeing of the hands, as noted by Dioscorides, is very interesting and could be implying intentional or unintentional uses. According to both Parkinson and Gerard, women would use alkanet as a cosmetic, specifically a rouge, that didn't last very long in France and other parts of Europe. This is in keeping with its use by ancient Greek women whom would create a stick rouge, "phykos", of alkanet for cheeks and lips. It was also used as a red hair colorant and pomade. Once again this can help to show why perhaps henna never became immensely poplar in Europe as they already had a native plant that worked just as good, if not better than henna. The 16th century herbalist and author, John Gerard, writes:

> "These herbs comprehend under the name of Anchusa, where so called of the Greek word that is, to color or paint anything. Whereupon those plants were called Anchusa, of that flourishing and bright red color which is in the root, even as red as pure and clear blood. The Gentlewomen of France do paint their faces with these roots, as it is said."

The alkanet plant is described as being weak looking and has bright blue flowers. Like a carrot plant, its root is quite large in comparison to the entire plant. It's also quite hairy and according to BODD, the leaves can cause skin irritation, hence why it was noted by certain writers of late that the leaves and flowers were never used…with the exception as salad greens and tea. The root is dark purple on the outside (rind) and grows ever lighter towards the center. It rots extremely quickly however and must be preserved via drying or eaten straight from the ground. There are said to be over 30 varieties which fall under the name of alkanet, some of which are thought native to the Mediterranean including Turkey, Israel and Palestine. Today, at least, it is found growing wild all over the Mediterranean, especially by water. This is not surprising as alkanet is extremely weed like. It spreads quickly and its seeds can withstand water travel. Research today has made connections with cancer and consumption of alkanet over a long period of time, causing some countries to ban its use in food or beverages. Safety of using it externally has also been called into question. In addition to coloring perfumes, food and wine, it was also used as a moder for cloth and a varnish for violins. Like "camphire", it is important not to make the mistake of combining the histories of these two very different plants. Through paying special attention to *its* history, I believe the mystery of why henna did not become popular in Europe is thus explained. This is seen by the very early 19th century writings of the German C. H. Ebermayer and other authors whom noted henna's only real use was to color things, such as oil (via the roots). They also remarked it was too *expensive* or could be adulterated, so Alkanet should be used instead.

THE MYRTLE MIX UP

> "The myrtle ensign of supreme command,
> Consigned to Venus by Melissa's hand,
> In myrtle shades often sings the happy swain,
> In myrtle shades despairing ghosts complain.
> The myrtle crowns the happy lovers' heads,
> The unhappy lovers' graves the myrtle spreads. -
> Soon must this spring, as you shall fix its doom,
> Adorn Philander's head, or grace his tomb.
> —Dr. Johnson

Myrtle (*Myrtus communis*) is our third major confusion causing botanical and we have India to thank this time, although their words have been totally taken out of context as you will soon see. As explained in chapter 1 and proceeding chapters, "Mehndi" really means myrtle but is also a name used for henna today. This most likely comes from the fact women of India (and most likely women of other areas as well) were staining their hands and feet with *other* botanicals, including crape myrtle, prior to the introduction of henna. People of India have long been likening plants to one another. Case in point, they call the kino tree "red sandalwood", even though it is not a type of sandalwood. Lack of Latinate names forced this sort of likening and in addition to myrtle, henna is likened to all sorts of staining botanicals including rue, baleria, etc., in India. Even though this is the case in India, it appears confined to the country and their ancient manner of plant classification. It does *not* mean that henna and myrtle were interchangeably used in other parts of the ancient Old World. Either by mistake or on purpose! This is very clearly seen by the lack of linking between the two plants by such ancient authors as Pliny Elder and Dioscorides. Some would very much like a "henna and myrtle" connection for esoteric and mythological reasons, something that henna lacks but myrtle drowns in. What some want and what is factual is two different things and history should not be denatured for it. In any event, the ancients were not dolts and took their offerings very seriously. They also took their plants very seriously, as a simple misidentification could mean death.

As noted many times in this chapter, Pliny Elder and others wrote about henna and other plants, including myrtle separately, making *no* linking between the two. They also describe myrtle, which was made into an unguent called Myrtinum, as growing in Egypt and being used to bind and thicken things. In ancient Egyptian texts, myrtle is believed to be "khet-des" and was used in numerous medicinal recipes. It was exclaimed by the ancients "the berries taste like wine!", which we know is certainly not true of henna berries which are unpalatable. The berries of myrtle were frequently used in medicines, condiments and wine for a pleasant sweet taste. Athenians were said to consider myrtle berries a confection with the added bonus of instant breath freshening. The berries were also a form of identification about when to begin harvesting the myrtle as well. According to Porta, Cato Elder (234 BCE), wrote in his book "On Farming" that myrtle twigs with their berries upon them must be taken from the bush while the berries were still tart. The leaves were then bound around the berries and this served to preserve them and allow them to ripen for culinary, medicinal &/or flavoring use.

Myrtle has a long history of high esteem among the ancients. Most likely because of the plants extremely fragrant leaves. Most true myrtle's contain oil in their shinny leaves which oozes out and creates a fragrant attraction, in addition to the flowers and berries which are also sweet smelling. Ancient Jews saw the fragrant leaves as a symbol of peace and also immortality as the leaves stay aromatic weeks after being picked from the bush. They also frequently incorporated it into their wedding ceremonies, as a symbol of a long marriage and in funerals, most likely as a way of covering the odor of the un-embalmed body. The Jews were said to eat the leaves of myrtle, which taste like a combination of rosemary and bay, in order to be protected from witches. It was also thought that if the myrtle leaf crinkled in the hand, ones true love would forever be faithful. Many other ancient cultures also revered the myrtle, including Athenian judges whom were said to wear wreaths of myrtle as a symbol of their power. Wearing a wreath of myrtle was extremely common, especially with the Greeks and later the Romans whom saw it as a symbol of passion, love and death. Even awarding them to Olympic champions in addition to playwrights and poets. Military heroes were given Corona Ovalis', also known as a myrtle ovation crown. Crowns were also frequently worn during weddings all throughout history, including the Middle Ages. This is expressed in a few lines by Drayton which reads:

> "The lover with the myrtle sprays
> Adorns his crisped tresses."

Myrtle crowns are some of the most compelling evidence that henna was not on the same level of use in ancient times. Due to the tight formation of myrtle leaves and the fact they stay small and compact no matter what the size of the bush/tree, makes them perfect for a crown. Henna's leaves are too big and also too sporadically bunched to be used for such purposes. Cato Elder even writes in his "On Farming" that people living close or in cities should grow botanicals for wreath making and suggests myrtle (not henna) as one to cultivate for such purposes. The crowns were not just worn for their esthetic purposes anyway, it was for their extreme aromatic nature as well. Further evidence of this comes from the finding, by archeologists, of preserved myrtle crowns / garlands from the Graeco-Roman period. Even more compelling is a myrtle wreath made of gold from the 336 BCE tomb of Philip the II from Vergina (housed at the Thessaloniki Archaeological Museum). It is simply amazing and so detailed, it looks as if liquid gold was poured over real myrtle leaves and flowers. It also gives a very good indication of how henna would not make a very similar or suitable crown or wreath. The flowers of the myrtle are large and very defined. The leaves are small and very close to one another on the branch. The branch itself is pliable enough to be bent into a crowning "V" shape.

Speaking of the letter V, myrtle is well known for being attributed to Venus. There are many stories about her and myrtle including one of her creation myths. When she emerged from the sea foam surrounding the island of Cyprus, she had no clothing and out of shame hid behind a great myrtle tree. She was forever grateful to the myrtle and it became a symbol of her. Archeologists have actually found age old myrtle growing around shrines and temples for Venus (also known as Aphrodite) on Cyprus. It was not uncommon for worshipers to plant special gardens containing plants devoted to the particular deity housed. Some of these ancient gardens still remain today.

Myrtle, like camphor and alkanet, had many medicinal uses as well. This is seen through the writings of Culpeper and others. Culpeper writers:

> "The Myrtle tree, the leaves are of a cold earthy quality, drying and binding, good for fluxes, spitting and vomiting of blood and stops the Fluor Albus and menses."

In the time of the 1st century AD and most likely prior, Latin texts by Soranus explains myrtle was used by women attempting to prevent pregnancy and in the Fragrant Garden, it is suggested as a wash for women suffering unpleasant odors of the privates. Cato Elder gives a detailed myrtle wine medicine recipe in his 234 BCE "On Farming" which describes myrtle berries being dried, then allowed to ferment and used to treat indigestion pains. Myrtle berries produce a non-alcoholic wine which was frequently used on children. Culpeper uses the leaves and berries of myrtle in numerous recipes such as syrups and ointments and a popular perfume was even made from myrtle and other ingredients in France called Angel water. According to a number of Victorian texts, the use of myrtle medicinally fell out of fashion and was rediscovered around the 1870's to treat a number of conditions. Most likely it only fell out of fashion in Europe but continued to be used extensively in other locals such as India.

The utmost evidence that the ancients were keen enough to know the difference and did not substitute henna for myrtle comes from a myth the ancients themselves created around the myrtle leaf. Myrtle leaves exude their extreme fragrance via numerous oil glands or pours located on each leaf. This gives the effect of tiny prick marks over the entire leaf. If you have the opportunity to examine a myrtle leaf, hold it in the light to discover this fact. The ancients too noted this and created the following myth:

> "Pheadra, wife of Theseus, fell in love with her step-son Hippolotus and when Hippolotus went to the arena to exercise his horses she would wait for him under a myrtle tree in Trozen. While waiting she would take a myrtle leaf and pierce it over and over with her hair-pin. The puncture marks in myrtle leaves today are a memento of her love."

This myth is also told at times in another manner, stating the tree was a meeting place for Hippolotus and Pheadra &/or Hippolotos was exercising for Olympic games instead of exercising his horses. In any event, this helps to bring home the point that the ancients were not as doltish as many modern people believe. Since making the mistake of picking the wrong plant either for food or medicine could be deadly, such stories helped to prevent such happenings. Thus, the ancients were extremely particular about the botanicals they used in conjunction to their deities and wouldn't have ventured to make substitutions. Since myrtle grew quite prevalently, there also wouldn't have been a need to make the substitution of henna.

CHAPTER THIRTEEN

❀

The Darker Side of Henna
Poisoning, Death and Other Unspeakable Things

> "Behind the veil of lighted candles which they held in their henna colored fingers, their lips smiled enticingly."
> —Nizami, Haft Peykar, 12 century

Like all things, henna has a darker side few know about or even wish to know about. Of course in the hands of humans practically anything can be used destructively but henna has inert properties which can encourage such happenings. One of these roles is as a poison and abortive substance. Being that women have seemingly known for ages henna had the ability to both prevent and discontinue pregnancies, is most likely the basis it is not truly seen as fertility inducing. Sensuality and fertility was most likely added to henna's cosmetic mystique much later in time (such as in the past 900 years). Henna was likely secretly used by many women in an attempt to abort babies, especially if they were prostitutes. Today this is seen in rural areas where poor women continued attempting to use henna for such purposes. This has not surprisingly led to numerous women having to be hospitalized. That is especially seen in Africa where almost half of young Nigerian girls and women (15 to 30) admitted to trying to use henna and alcohol to abort unwanted children and later found themselves very ill. Numerous women have died from such tactics. Documented use of henna as an abortifacient has been made in Africa so obviously many women were successful. Likewise a number of new born babies throughout Africa and the Middle East have perished due to being dipped in henna. Like the "cooking fire" excuse one has to wonder if this can in some circumstances be female infanticide which is *know* to occur in such areas. Henna is documented as a poison in Mexico, thus leaving one to wonder how many people have died as a result of internally using *Lawsonia*. Thus, henna has been incorporated into rituals done at death or more gruesomely before an impending demise. Although tales of Moroccans hennaing their captives is somewhat dark, imagine decorating your hands with henna in preparation for being buried alive. This is certainly what a number of women did in Rajasthan India which was called "Suttee". One such story that is recounted is of a woman who sends her husband off to war. When she hears that he was killed in battle she calls for all of her servants and instructs them to prepare henna as they would for a wedding. The servant women reluctantly did so. They then, while sobbing at the impending demise, elaborately decorated her feet, legs, hands and arms with henna. She remained composed and told them it was her duty to perform suttee. A number of poets also wrote about women's use of Mehndi prior to committing suttee. Sekhavat wrote in the voice of a woman, "Decorate my feet with Mehndi rich, O Nayan, my Lord has gone to battle. When he comes home victorious I'll celebrate, If he dies I'll perform suttee and settle". Malla Misran wrote, "Nayan do not deck my feet today, I hear tomorrow will be a fray. You'll deck my feet with Mehndi then, If my Lord gets killed in the affray." Such stories and poems are confirmed by a particular fort wall in Rajasthan from the 15th century. On the white stone walls of Fort Bikaner you will see hundreds of small, red hand prints. These hand prints are all of women who committed suicide on their husbands funeral pyre's. Legend has it that during the funeral the women would dip their hands in henna paste and then press it onto the wall to leave a legacy behind. Later, masons would come along and cut out around the hand print to make them permanently etched into the wall. Women who committed suttee were seen almost in a saintly light. In the beginning it was the richer women who performed suttee as they saw after their Maharajah died, they would no longer have a high place in society. All sorts of romanticized writings started and women of all classes began to join in. It was later banned because of the popularity that ensued. The famous Indian poet Tagore, who was educated in England, became a great voice in the effort to stop such senseless suicides. One can somewhat understand losing a spouse and feeling so lost that suicide becomes the only escape from the pain. However, suttee was intrinsically hinged on the woman's new found worthlessness in society. Family members would encourage the suicide for their own selfish reasons, such as getting a hold of the estate and the sainthood marker was also deviously appealing. This is especially seen by the lack of male suttee participants. Women of the area were also known to build huge fires and collectively commit suttee when they knew their camp was being invaded. They were said to rather die than be captured by the Moguls or other invaders. So comes in yet another awful reality about henna and women in India. Cooking fire deaths. After seeing pictures of charred and burned bodies, still dressed in richly colored saris and henna decorations, which I will never forget, the luster and romance of an Indian wedding is simply no longer there for me. So happy are the brides decorated with henna, one organization states which is trying to bring light to the horrific acts. Little do they know that once they walk through their grooms doorway they will be doused with gasoline and lit like a torch. It's so ridiculous the excuse made for the brides horrible demise, a cooking fire. Everyone knows a bride does no work until her henna fades. Yet the cooking fire is repeatedly blamed and no one *ever* goes to jail for the murders. Even worse is the horrid crimes are many times planned and carried out by the female members of the household. They want money, specifically dowries which are supposed to be banned in India. Over and over again they murder their sons new brides until they amass the wealth they think they deserve. In one report by UNICEF, 5,000 women a year were estimated to have died in India alone as a result of bride murders. This number is said to have gone up significantly. This simply tarnishes all that is romantic and exotic about Indian weddings, including the use of elaborate henna designs. However, henna can and should be used to bring light to this ultimate form of gender bias and abuse to women. If one performs Mehndi for a bride to be, why not donate a portion to a charity that is working to stop the senseless murders taking place today? Henna's popularity in the US climaxed in 1998 and it no longer is getting any air time. In a sense, the air time surrounding henna was so horribly wasted with sugar coated ramblings of sensuality and romanticism that one wonders if it can ever be used again for real *good*. Yes this is gritty and disturbing but it's also real. The public deserves to know everything about henna, not just the exotic and appealing side. The one good thing about Jogendra Saksena's "Art of Rajasthan" is the way suttee and all things dark were included in conjunction to henna's history. All be it romanticized but still included. In the light of today's fascination with henna as a feminist platform and perfect example of a *fine art* created by women, I find it very disturbing that mention is always neglected by Western writers about henna's darker side. Why is it so taboo? This leads to yet another dark realm many don't wish to speak about. FGM or female genital mutilation. Some call it female circumcision (or excision) but in reality it is much closer to the prior mentioned and will be so called here. It has been banned in many of the countries, including India where it is practiced, which has only driven women underground in their quest to keep the procedure alive. Henna unfortunately has long played a number of roles in FGM. In some areas including Morocco, the women who would traditionally perform henna body art (hennayat) and midwife (Daya, Gedda) services would also engage in FGM for extra income. The henna was used as a suture like glue to help hold together the incisions and promote healing. It was also used to try and stop the immense blood loss and as an attempt to clean and serialize cutting implements which included knives and shards of glass. Likely, it was also employed to stop the odor of putrefying tissue, blood and urine which is all too common. FGM leads to horrible birthing experiences and life long pain. Some Westerners, for some unknown reason, have rallied around the practice stating it is a women's form of sensuality and body modification. They exclaim, women should be allowed to do what they wish to their bodies. This is true, *women* can do what they please but the actual recipients of FGM are young girls. Not women. Sometimes as young as 3 years old. The average age among all who practice it is 5 to 8 years of age. They are never asked, "do you want to go through with this?" but instead forced into it by their mother and relatives. They are not told how it could effect their fertility or how they may die because of all the blood loss. FGM is certainly a form of child abuse. The reasons their mothers continue with the practice is the same as why tattoo artisans don't use numbing agents today. They could if they really wanted to but when they got their tattoos it hurt. They feel that their patrons should endure that same pain as well. This is what drives the FGM to continue, a "if it happened to me and I lived, it will happen to you" mentality. Ignorance is also a great factor. Parents think that girls will be unfaithful to their husbands without the FGM or that they will not be able to prove their daughter is a virgin. Many organizations have been started (many times by women who went through FGM) to help stop this practice and hopefully like the foot braking and binding of China, it will soon be eradicated. Likewise, henna was also used in male circumcisions quite prevalently. Much for the same reason as women, to help with healing and stop any blood loss. This has led many men to have a horrible memory of henna and great dislike of its aroma. This is especially true of men from the Middle East. In some more modern areas, the henna use resembles how it is used in weddings, instead of it being medicinally applied to the baby or child. Male circumcisions are seen as a happy occasion and henna is used in Jewish and other such ceremonies.

HENNA'S SEEDIER SIDE

Many have noted that henna is frequently depicted on hands and feet of prostitutes and street workers. This has even been photographed in various areas all around the Old World. Specifically in the Near East, Middle East and Africa. This however doesn't mean that henna was reserved for such women, which are frequently looked down upon. Even today, many of these women are noted as wearing henna, even in modern areas where it has gone out of style. As a number of people have written however, their use is similar to the use of all women in their particular area. Usually, they seemed to have a bit more money (and time on their hands) than the average house wife, which would allow for splurges on henna no doubt. Not only that but looking exotic was important for such women, just as it is today for women in the same situation. Everything is exaggerated, and likely this was so true of henna body art on such women as well. Henna also had another purpose, still believed to be true today. That is a protector from STD's and venereal diseases. It is likely that, originally, a driving force of henna use among such women (and, sadly men too) was its protecting value. Such a protective value was even thought to inhabit the stains left behind by the henna paste. Preventative use of henna likely always led to treatment of the actual STD by these same women again using henna. Herpes in the mouth (and other areas) was frequently treated with henna. This continues even today. Henna has been found to have very good fungal killing properties and also the ability to kill germs but it has a dramatically less effective role on viruses. It would be very unlikely henna could prevent the transmission of STD's. Likely it would only mask the sores and other markers of the outward infection. Prostitution in the Middle East and elsewhere was not reserved only for women. Many brothels were inhabited by men who seemed to also use henna, likely for the same purposes as the women, as noted by Burton who wrote:

> "The Afghans are commercial travelers and each caravan is accompanied by a number of boys and lads with khol'd eyes, rouged cheeks and henna stained fingers and toes. They are called kuch-i-safari or traveling wives."

In the Islamic religion, men are specifically forbidden from using henna on their hands or feet in a manner that women did. This apparently didn't stop many such men of the night from doing so. Some Victorian authors also noted boys cross dressing in women's Oriental dance attire and doing the same dances as women. They also completed the look by using henna decorations on their hands and feet. William Lane wrote:

> "Their (the Khawals) general appearance, however, is more feminine than masculine. They allow the hair of the head to grow long and generally braid it, in the manner of the women and they imitate the women also in applying kohl and henna to their eyes and hands."

This was likely going on for a very long time in the Middle and Near East and may have been what caused such restrictions on henna being used on men. Or at least reinforced the notion of henna used on the hands and feet (decoratively) as being very feminine. Most of these Khawals were young boys who were exempt from the Islamic "no henna" rule.

FURTHER INTO THE SHADOWS

Many Orientalist writers and today new age writers try (and tried) to place a heavy esoteric weight on henna. This was most likely so, especially in the case of Orientalists, because they did not quite understand why people were staining themselves foreign colors. Anything exotically different was cast into the realm of witchcraft. Eventhough it had nothing to do with such uses. One can also think back to the notations made by Michaelis about people supposedly burning the henna designs into their hands and how evil can be attributed to that quite easily. The key to this is the wrong information which is then modified and finds itself in many books being sold today. For example Homer (Man and His Gods) wrote about the Egyptians seemingly not using human blood scarifications. He writes however that they made up for this by using henna:

> "The people of the Nile continued in the magic of red. They tinctured their nails with henna extracted from the privet and continuously applied red ocher in the form of carmine paste to the face and body."

From this, it is likely a great deal of incorrect information was extracted. Such is the case of henna being used in Goddess cult ceremonies. There is no evidence that the Egyptians used henna for purely magical reasons and when you read further on down the chapter, Homer starts speaking of the Egyptians literally having the ability to cut off and reattach severed snake heads without killing them and so forth. It casts doubts as to if he really knew what he was talking about. So is the case of henna being used in other types of spells. One is henna powder being placed in the shoes of some Muslim brides, forwhich she would perform special dances over knotted ropes. Then she poured the henna out onto the ground. This was supposed to keep away the Evil Eye and preserve her virginity prior to getting married. Obviously the henna was an after thought and the knotted items were the main portions of the *magic*. According to one antique book, Muslims were (and perhaps still are) allowed to use spells as long as they do not involve or call on the name of God. In any event, this is how one can see henna being throw into the mix simply because it is there. Hence how henna is included in some types of trance dances in Africa, specifically Morocco and the Sudan. Since women wear henna *anyway*, it's easy come to the conclusion that the henna places no real role in these ceremonies. What ever is the main focal point of the evening, is in fact the oldest form of the ritual. In this case it would be the dancing. Commonly called Zar trance dancing cults or "healing dances", the whole bases of the dances are evil spirits called "jinn". There is nothing healing about the dances, especially for the animals sacrificed, as the main focus is being possessed. Many agree that these cults hearken back to pre-Islamic African (voodoo) ceremonies. Ceremonies that were practiced by ancient non-henna using Africans. This however is likely what cast henna into the category of being an herb used in exorcisms even though that is not really correct. The Zar cults also don't seem to have a long period of structure as mention of them is made around the 1860's. A number of countries have outlawed the dances but this has simply driven them underground.

More bizarre henna history comes from a sickly little Agustinian nun named Anne Catherine Emmerich. A stigmata sufferer, she also was known to have visions of the past and present. In 1813 she was removed from the church housing and was forced to live with a woman, where she was confined to the bed. This is where she began asking the poet Klemens Brentano to record her visions. He later published these visions in three books in the 1850's, after her death. The henna visions are found in the second book, "Life of the Blessed Virgin According to the Meditations of Anne Catherine Emmerich". In one of her visions of Mary and Joseph traveling during the crucifixion she recounted the following:

> "I particularly noticed a plant which had pretty small green leaves and branches of flowers, made of nine closed pink small bells. Perhaps this flower with nine bells had for the sister a mystical value. The flowers indicated here was probably the small branches of cypri (*Lawsonia spinosa inermis, Linn.*) of which it is known in the Songs of Songs."

She then went on to say the flowers were fragrant and compared them to how Eastern women used henna for personal perfuming at her time. This she said is how Mary or Elisabeth used henna because it's perfume was something a pious woman would use (obviously the visions didn't allow for the sense of small). Then she went about quoting that this cypri was henna in the Songs of Solomon as well. The remainder of the henna notes were in relation to comparing henna to blood and death. Specifically that the bunches of henna was like bunches of blood. Who knows if this is what prompted the supposed curse Gypsy people feel henna has over it. Specifically that Mary used it before Jesus died (making the assumption the *henna* was the cause of the death in some way). Emmerich's visions sold vast copies all over Europe in the mid 1800's, including Spain so this could have been a great possibility. What may be actually seen however is that Christianity of any sort didn't seem to ever condem henna. It is likely the notations of Giovanni Mariti (1761) from his travels in Syria and Palestine had a profound effect on Emmerich and much of the Catholic would, as he too noted Eastern women used henna to perfume themselves. Emmerich's description of henna flowers doesn't seem all to correct however, leaving one to wonder if the constant reading of books had a profound effect on her visions.

Chapter Fourteen

❀

Vineyards of En Gedi
The Cultivation and Harvesting of Henna

> "His garden was full of trees and was well fenced around with a ditch and henna hedge."
> —Mrs. Sherwoods Stories, 1817

This chapter will hopefully accomplish a number of things. Number 1, this will be of great use to anyone who wishes to grow their own henna for personal use. According to a dye crop book published by a department of the UN in Rome, the availability of published agricultural information on henna is *poor*. Until now (with this book), the majority of agricultural information about henna has been spread about in bits and pieces. The bulk is also from Indian sources predating the 1950's. This chapter will combine growing information from all of the major henna producing locals in addition to cultivation in Victorian England and my own *personal* insights into growing henna here on the East Coast of the US. This chapter will also be of great use to farmers of any size wishing to obtain information about growing henna for commercial purposes. I will however be neglecting to include information about the dangerous pesticides many growers use, as I believe henna should be grown in an organic nature. Henna plants require care but, according to my experiences, are not as susceptible to pest damage as many suggest. Growing henna is not as hard as you think and it is really adaptable to various climates and conditions. If women in Victorian England and the mountains of the Caucasus can grow it, you can too. If you would like to grow henna on a small scale (it makes a great tub plant), I suggest buying cuttings which are already situated in soil and ready to transplant. For a company that sells wonderful small henna plant cuttings, skip over to the Appendix of this book. Larger growers will want to use seeds which may be found at many exotic nursery plant sources. Look on the Internet search engines Yahoo!(TM) and Altavista.com using the search words "Lawsonia inermis or alba seeds" and a number of companies will pull up.

BOTANY OF THE HENNA PLANT

> *"Alhenna (Lawsonia): A genus of the Octandria Mongynia class. The calix consists of 4 segments and the corolla of 4 petals. The stamina are disposed in pairs and there are 4 capsules containing a great many feeds."*
>
> —*Encyclopedia Britannica 1771*

Division: Magnoliophyta
Class: Magnoliopsida
Sub-Class: Rosidae (Cronquist's Subclass Rosidae)
Order: Myrtales (Dahlgren's Superorder Myrtiflorae)
Group: Dicot (Dicotyledonae)
Faimly: Lythraceae (Lawsoniaceae J. G. Agardh, Lawsonieae, LAWS, Lythraceae J. St.-Hil)

Main Species: *Lawsonia inermis Lamarck, L. alba., L. ruba, L. spinosa* and *L. mimata* [?].

Henna is basically a glorified hedge. If not for its coloring abilities and extremely fragrant flowers, it's doubtful it would be used for anything but hedge purposes. Thus, henna is considered to be deciduous and an annual, in addition to an evergreen in very hot locals and even a perennial at times. It is also non-succulent. It is a member of the Lythraceae family which is commonly known as Loosestrife and includes other plants such as the crape myrtle. In addition to Lythraceae J. St.-Hil, in the mid 1800's henna was placed under the family title of Lawsoniaceae J. G. Agardh, where it is the only member. It appears from records that the henna plant itself was first given the Latin name of *Lawsonia* in 1753. In 1763 it was then placed in the family of Lythraceae. Then it was placed into the very particular family of Lawsoniaceae or Lawsonieae. Henna, if maintained, can be kept cut small and bushy but if allowed to grow freely, can reach heights of over 20 feet tall. The bark of henna is gray, brown in color and is smooth, with no spines of any sort on *Lawsonia inermis*. Other species (such as *spinosa*) may have thorns at the base. It is considered glabrous and has many branches. The leaves of the henna plant can vary a great deal, even on the same plant. From my experiences with growing henna, when the leaves are near the end of their lives they begin to curl and are quite long, with slightly serrated edges. When they are young they are quite ovate, smooth and flat. Leaves directly under the flowers always stay small and young looking. Henna leaves are medium green in color, elliptic or broadly lanceolate and soft to the touch. When broken, a yellow juice may ooze out and they have no scent in fresh form. They grow oppositely paired with one another and sometimes in a whirled manner. Many liken the leaves (from a distance) to that of myrtle or olive. Their normal size is 2 to 3 cm long and 2 cm wide. They are either xerophytic or helophythic rooted. Lamina margins are entire and leaves are without a persistent basal meristem. The stems have cork cambium present, initially being deep-seated. The Primary vascular tissue is bicollateral and an internal phloem is present. Secondary thickening develops from a conventional cambial ring. Vessel end walls are simple and the vessel's themselves have vestured pits. The flowers are quite small (about 1/4 of an inch) and grow in panicles or grape formations (as lilacs do). Extremely fragrant, they come in a number of shades including (most popularly) white and yellow as well as pink, red with shades of purple also being found. Many find that the white flowers turn ever yellow in color as they age. On extremely close inspection of a single flower, one will see it has 4 thick petals (or 4 merous) which are very crinkled (when fully cyclic, usually pentacyclic) and attached to the inside of the floral tube. Free hypanthium is present. They also have a simple style and superior overy. The calyx is bowl-shaped, without appendages and may have a central vein running through it. Stamens grow in pairs, equaling 8 and are quite long compared to the flower as a whole. This is likely so because henna is a "hermaphrodite", meaning it is self pollinating. It does not need animals or insects to perform the pollination and reproductive process. Pollen grains are 3 or 9 aperturate. The berries which come after the flowers die, are blue-black in color and non-edible. They are also non-endospermic. Commonly called an indehiscent type of berry, the actual seeds found inside are extremely small (the size of poppy seeds) and triangular in shape. Their outer shell is extremely hard, with a spongy tip and allows for travel in water before germinating. C3. In general, germination is phanerocotylar. After seeing and experiencing live henna plants, one can get a better appreciation of how Dioscorides' description was very on the mark. If you'll remember, he noted henna as having full branches of olive like leaves except they were larger and softer to the touch, white pleasantly scented flowers and black seeds. Henna is believed to have originated in North Africa but today is found growing wild all over the world. This includes, but is not limited to, Egypt (and other parts of Africa), the Middle and Near East, Southern China, Europe, India, Indonesia, Java, Malaya, the Philippines, Mexico, Dominican Republic, Cuba, Madagascar, Haiti, Australia as well as many of the Caribbean spice producing islands. It is considered a temperate or topical loving shrub. Today it is noted as a cosmopolitan, except in very frigid areas (X=5-11). Some also say that a single species of henna has Palaeotropical distribution. Likely this is *L. inermis*. The Palaeotropical section of the Old World is however very broad (spreading from Africa to Polynesia) and henna must be examined more closely to determine the specific floristic regions and sub-kingdom's it actually belongs to. So far this apparently has not been scientifically done. Henna being Palaeotropically distributed somewhat differs from the notations made by many about it originating in South North Africa (Morocco, etc.) as that is considered Boreal. In addition, Indo-Malaysia is a region of the Palaeotropic, yet henna grows better in the dryer "Boreal" areas of India instead of its tropical band (comprising Bombay). This is likely because henna needs very specific growing conditions not found in all tropical settings. Like the conditions found around the Nile and its tributaries. The Palaeotropical floristic kingdom *does* include Ethiopia and its White and Blue Nile's which is connected to the real Nile of Egypt. Personally I have yet to see any research done on henna growth in Africa (likely for reasons I mentioned in chapter 2). It would certainly be beneficial for botanists to start researching henna growing in both Ethiopia and the Sudan (especially the area of ancient Nubia) so that its true origins may be found. Since coffee is believed to have originated in Ethiopia, it isn't very much of a stretch that perhaps henna first appeared there as well. Henna could have even migrated before people moved into the area, specifically via the Blue Nile, into the Nile proper, which deposited it into Egypt (especially about the Delta). This would correspond to how henna appeared to be mass grown in upper Egypt and the oldest literal artifacts found there as well, which was located at Hierakonopolis. This is more plausible then other theories (predominately consisting of human introduction) when you research how both the White and Blue Nile's travel. As the White and Blue Nile's separately work their way through Ethiopia and into the Sudan, they come together to form the Nile proper of Egypt. As the Nile runs its course through Egypt, it splits into tributaries forming the Nile Delta which intern flows into the Mediterranean Sea. Since henna is self pollinating (with no known insects or animals aiding in the process) and the seeds so extremely hardy (in that they must be soaked in water to facilitate germination), water travel seems perfect as a mode of migration. Both Sir Samuel Baker and Amelia Edwards noted they saw henna growing in profusion on the Nile tributaries of Abyssinia. Edwards also noted, in her Victorian era, that the Nile tributaries in the Sudan (specifically Nubia) was the main provider of henna to the entire of Egypt. It is likely henna was waiting for the ancient Egyptians even before they arrived due to natural migration of its seeds. When you compare all of this information with theories of henna coming from the Mediterranean or Morocco to ancient Egypt, the latter appears rather absurd. In addition this would explain why henna was seemingly not introduced outside of Egypt until the exudes of the Israelites or later. Similar to how coffee was not used until much later in Old World history. Coffee and henna differ however in that the latter could use water passages to spread to new locations where as the *Coffea* always remained in Ethiopia. Once again, it would be prudent of botanists to focus on henna growing in Ethiopia, the Sudan and around the entire length of the Nile and its tributaries.

In Morocco (as well as other areas) a number of types of henna are recognized. El-ghafeky has leaves that resemble that of the pepper plant, being long and thin. The flowers are extremely fragrant and red in color. El-madjoussy (or El-madjoun-hinna) has varying shades of leaves on the same bush. This can include a range of taupe to jade green. The flowers also can range a number of colors including white, yellow and purple. Just before they expire, they produce a cotton like fiber. This type of henna is frequently seen in the mountains. Maghrebin-hinna has the darkest of all the henna species in the area and densely packed pinnacles of white flowers. Species called El-koreiche and Ouesma are seen as having great medicinal worth. *Lawsonia inermis* is seen as true "hinna" and its white flowers are known to be used to make the actual yellow to red dye used for body art in some areas. Hence why it is called "fingernail flower" in China and Japan. In India *L. inermis* is typically seen as being red henna and superior to *L. alba* which is commonly called black henna. In India, red henna (*L. Inermis*) appears to be predominately used for dye while black (*L. alba*) is used for perfume purposes. Some note species such as *Ammania auriculata* as being a type of henna plant. This however is incorrect, the above mentioned and other *Ammania spp.'s* are really simply in the same family as henna, Lythraceae. There are hundreds of plants that fall under Lythraceae but only Lawsonia falls under the family of Lawsoniaceae. Therefore subspecies would be *ruba*, *spinosa*, etc. Many of the other species, especially in the *Ammania* family are blistering agents when applied to the skin. They also do not resemble henna bushes, so it is unlikely people have been confusing the species.

Due to henna's extremely bitter taste (and sometimes thorns), it does not facilitate grazing by animals. It was also found in studies to be so low in nutrition that animals were not able to sustain themselves on it alone, no matter how much they consumed. Thus, it was deemed unsuitable as feed. Therefore it was likely many herders and those living in very rough conditions found henna to be worthless. In some areas, henna is considered a pesky, introduced weed. So, its usefulness is totally based on the peoples needs and complacency in its immediate area. In that way comparable to the rose.

HENNA IN THE GARDEN

> "*It is the moonlight night of March; the sweet smell of henna is in the air; my flute lies on the earth neglected and your garland of flowers is unfinished.*"
>
> —Rabindra Nath Tangore, The Gardener

As in the passage in Mrs. Sherwoods Stories, henna has been a staple of the "well-to-do" garden setting. Many historians believe henna was a plant grown in ancient Egyptian estate gardens along with other extremely fragrant plants. It is also seen in the estates of Jericho and King Solomon, although it is believed henna was introduced to these areas. Like orchids are to the rich of the 20th century, henna appears to have been a plant reserved for those with prestige to grow in their huge gardens. In addition to providing extremely fragrant flowers, the henna, when cut correctly could form a nice hedge and leaves could be harvested at will. Nubia, well into the Victorian period was a major producer and site to experience wild henna. Especially along the Nile. Baker wrote that he saw henna growing in considerable quantities along the Nile and Amelia Edwards noted the same on her trip to Nubia in the 19th century. She also noted that on her trip to an ancient temple in Nubia, located specifically at Gef Hossayn, henna bushes, dwarf palms and acacia grew all about the ruins. Edwards likewise observed that at the time of her visit, Nubia was a major producer of henna for Egypt. In Victorian times, the henna was turned into fancy potted topiary plants called "Mignonette Trees". This would include having a bare trunk and a sphere of leaves at the top which would be covered with fragrant pentacles of henna flowers. In cooler weather, the bush would be kept indoors. Thus, a fancy tub plant. Henna was frequently grown near rose bushes, as they are wonderful companion plants which aid in deterring pests and diseases from one another. The scent when both were in bloom must have been marvelous to behold. Growing henna in ones garden is still being practiced in many areas of the world including the US. Peoples in Arizona and New Mexico have found it quite easy to maintain a lovely and fragrant henna bush outdoors year-round. The most well suited climate for the henna plant in the US would be the tip of Florida. Other areas will find themselves needing to take the henna indoors when the temperature falls below 19 degrees centigrade or covering it with a black plastic bag to prevent frost damage. Henna is indicated for continuous outdoor growth in zones 10 and 12. Plant hardiness zones created by the US Agricultural Dept. has now been adopted by many areas of the world. Look at an agricultural map of your area to find out what your hardiness zone is. Areas such as Mexico here in the Americas would be considered 10 and are perfect conditions for growing henna. Don't fret if you don't live in such hot zones, simply grow your henna as a tub plant indoors.

HOW TO CULTIVATE HENNA

"Henna's hardiness zones are 10 through 12."

—Agricultural Zone Chart

Growing your own henna is *not* as hard has you think or imagine! Many attempt to deter growing henna oneself by throwing all sorts of complexities at you. Your henna needs constant extreme heat, the henna is susceptible to insect damage, you can never grow henna in the US or in colder climates. Well, speaking from experience, this is all very untrue! Remember henna now grows in China and on the Caucasus mountain tops. It's very adaptable if it's treated right. The biggest mistake that is easily made is the failure to provide it with enough water! This is natural because you figure it is a desert type plant and you see the rainfall needed is only 0.2 to 4.2 meters a year and so forth. Don't be stingy, give your plant water when the soil is dry, such as every other day. Don't deluge your plants and cause root rot, but don't let the leaves dry out either. Leaf fall will ruin your chances for a nice henna harvest. First, lets start at the soil and work our way up. Henna can grow in all sorts of soil conditions including sandy and stony, depleted dirt but as many plants do, it grows best in rich soil or better yet loam with fertilizer added. What is wanted is a nice water retentive growing medium such as medium sandy loam. This will recreate growing conditions near the Nile, where henna grows best. While henna likes moist soil, it doesn't like a whole lot of moisture in the air. Henna is considered a sub-tropical plant which many find grows best in dryer areas. In India it's most popularly grown in Rajasthan instead of areas like Bengal which is too wet. If your property or land tends to flood a lot, henna may not be suited for you. Growing it as a tub plant may be better. Using fertilizer is suggested and compost or fish extract may be used. Always be careful not to burn the roots of your plant! If you have old fish bones (or heads) laying around you can make a sun-tea out of it, dilute it with more water and use it every time you hydrate your plant. Be very careful using compost from your own garden however as it can be full of disease that your henna plant is not immune too. It can also harbor parasites and viruses. If you need to use compost, purchase some from the garden supply store. Many find that iron in the soil helps the henna leaves to dye brighter in color. With my plants I ground up iron supplements and added the powder to the soil. For larger quantities of additive look in your local garden supply store. The soil pH should be around 4.3 to 8.0. Not having such conditions may cause leaf drop.

Next you will need to know how to prepare the seeds or cuttings. For those new to growing exotic plants or if you wish to grow henna as a tub plant, try to obtain cuttings of henna. Henna is wonderful because you can snip of a branch and create more plants through cuttings, which I will explain how to do a bit later on. For larger farming of henna, seeds will be more economical. Henna seeds have very tough shells. This allows them to be carried in water for long periods of time before being deposited on land and quickly germinating. Due to the seeds nature, many believe (including myself) henna was carried to various areas via water, such as the Nile and even the sea, as salt water is unable to effect the germination capabilities. This would also explain why henna is frequently found growing near water. In any event, many find the need to soak the henna seeds in water over night or for 24 hours before planting them. Just as you would for bean sprouts and sprouts from lentils which two come in hard shells when dried. Another way to encourage sprouting is to place the seeds in a very cool place, many suggest the refrigerator. To do this, take a try and lay cloth or paper on top. Spread the seeds over the paper and place them into the refrigerator. The paper used can also be treated with gibberellic acid (GA-3) to help in sprouting. They can sit there for about 2 days before planting. In some areas of the Middle East, henna seeds are soaked in water for a whole week and then placed in baskets until they sprout. You may also have to split the hard outer coating of the henna seed to facilitate germination which is most commonly called "chitting". Once your seeds are cooled or have soaked or have been chitted, you can plant them in small trays or directly outdoors. You can also allow them to sit under a damp cloth (treated with GA-3) until they begin to sprout and then plant them. Henna seeds germinate rather quickly, around 8 to 15 days and then can be weeded out to make proper spacing, so don't worry about spacing in the beginning. A 20% germination rate has also been noted so don't be stingy on the amount of seeds you use. The more seeds you plant, the better chance of getting a few healthy plants. If planting outdoors, make sure your seeds or sprouts are in a direct sun setting and you may wish to space them 20 to 30 cm's from one another. After sowing sprouts, some will cover them with manure. Henna requires full sun which means it needs a good 4 to 5 hours of direct, hot sunlight in order to grow properly. If you are growing henna indoors or as a tub plant, you can use a grow light for the 4 to 5 hours of sunlight needed. Likewise henna should not be grown under larger trees or shrubbery. Some also recommend flooding the planted sprouts with a good deal of water while they are young and then when they are a few inches tall, let them be watered via rain fall. One must be diligent to pull up weeds around the base. Such weeds can rob your plant of valuable nutrients from the soil and introduce pests. Planting companion plants however will help your henna thrive. You can also place newspaper soaked in water around the base and cover it with soil. Plants will typically grow 20 to 30 cm in the span of 3 months. Even if you are not planting your henna in the ground, you can place it in a pot / tub next to companion plants. Each plant has a number of plants specific to its nature which play off of one another and helps to keep each other healthy from pests and viruses. This is many times called husbandry. Not surprisingly roses (*Rosa spp.*) are one of these companion plants. If you have a rose bush, nuzzle you henna plant next to it, they will both thank you for it. Other flowering plants include lupines and marigolds for pest control. Companion herbs include chives, garlic, basil, hyssop and parsley. If you have a big enough tub, you can even plant any of the above mentioned herbs or flowers at the base of your established henna plant. As for watering, try the orchid method of letting the soil dry out before watering again. Make sure you give them enough water however. If planting outdoors and you get a normal rainfall of 0.2 to 4.2 meters of rain a year, then you will hardly have to artificially water your plant. Try to keep your plants temperature at 19 to 27 degrees centigrade at all times. A constant temperature of 65 to 70 Fahrenheit indoors is very comfortable for your plants. Lower temperatures will encourage leaf drop. Henna plants grow *very* quickly when properly cared for. Many farmers are able to have a nice small harvest within 12 months of planting the henna. Of course what would seem small to a farmer would last you or me perhaps 6 months. On a small scale henna can be harvested when ever you need some. On a large scale, henna produces the most in its 2nd and 3rd year of growth. Many henna plants live about 4 to 6 years when harvested heavily. Although a number of farmers maintain henna for 12 or even 25 years before planting more. It's all up to your needs. Older henna can produce abundant flowers and bark, while smaller plants produce henna leaves. In henna's first year it should be pruned (as high as around 20 centimeters) to keep it low and bushy. The main time to harvest leaves is in the Spring time. Spring and fall are the main times to harvest henna in general. Fall is also a time to arrogate the soil and propagate. Henna should be watered often and allowed to have two dry periods a year. In this time the henna can be pruned and harvested as well as cuttings made. Always keep an eye out for pests. Henna is very resistant to pests because of its tannin and other inherent chemicals. This was noted by a number of *modern* growth researchers including the "Central Research Institute for Dryland Agriculture" or CRIDA (India). In fact some researchers were isolating natural chemicals found in henna for use as pesticides on resistant crop destroying insects. This is because Lawsone is one of the most naturally occurring phytotoxic phenolic compounds available. Blasting it with pesticides is very foolish, especially if you will be turning the leaves into paste and leaving it on your skin for hours. Henna's main pests are the dreaded henna wasp, termites, ants and mites. CRIDA noted it is not grazed upon by animals, because of the awful taste of the leaves, so you don't need to worry about that. Remember too that garlic can be a companion to henna and it also makes for an extremely good pesticide, not only for bugs but for larger animals as well such as rabbits. It is very easy to create your own garlic pest control wash, which will not harm your henna plants or you. Follow this recipe:

Garlic Pest Control Wash
4 Cups pure water
1 Tablespoon fresh minced garlic
1 teaspoon garlic powder (optional)
Blend all ingredients well in a blender and bring to a rolling boil. Allow to cool and place in a spray bottle. Spray entire plant as often as you need.

Likewise you can also use essential oil of hyssop and basil around the base of your henna plant or create a wash out of them. I have personally used such a wash on my henna, roses and other plants with great success. The smell as well as the taste is extremely repulsive to all manner of pests. To prevent the garlic from intruding on the lovely smell of the blooming henna, try using the basil or hyssop instead. To keep ants at bay, use ground black pepper. There are a great number of organic pest control products on the market now as well. Although you will not be ingesting the henna, it is smart to buy products that would be suitable to be used on vegetables and fruits to prevent toxic problems later on.

Another problem for many is actually harvesting the henna leaves from their plants. This is where maturity comes into play. Your henna plant needs at least 1 full year of growth before you can harvest just a small amount. Traditionally, henna is picked during the day and on a very low humidity occasion. Picking henna on a wet day can hinder drying and allow mold to creep in and ruin all of your hard work. While it sounds drastic, you will need to cut off an entire branch, not simply pick a leaf here or there. The leaves will not grow back and your henna bush will look a mess. Taking the entire branch off is the best way...you'll need to prune your bush anyway. Spring and fall are the traditional times to do this or after the henna has flowered. Take hedge clippers or a sharp knife, which has been sterilized, and cut the branch off cleanly. Leave the leaves attached to the branch. Set your oven on as low as it will go. Lay the branches (brake them into manageable sizes such as a foot a piece if need be) on cookie sheets and place them into the oven. Now traditionally the branches would be simply laid on the ground and dried in the sun, after which the leaves were striped off, but the oven technique will prevent the bleaching of the green color of the leaves and also the chances of mold. Henna needs to be dried in order to produce dye for the skin, fresh leaves do not work and are a myth from India that they in fact dye vibrant red, black or any other color. This is explained in chapters 7 and 9. Don't try to dry your leaves using a microwave. Microwaves cook the leaves and bombard them with radiation. This may negatively effect the medicinal and dying ability of the leaves. Your goal is to dry the leaves, not cook them. Once the leaves are dried, you can do with them what you please. They should of course be picked / striped from the branches. Many wish to grind them into a fine powder. A coffee grinder can work very well for such purposes. For more on this read chapter 9. Traditionally in China henna flowers are picked in the Spring. Depending on your climate and circumstances, you can make a crude perfumed oil from the flowers as Pliny and others did. Distillation works very poorly to properly extract the true scent from henna flowers, so instead try this method:

Henna Flower Oil
2 Cups vegetable or sweet almond oil
Henna flowers crushed (as much as you can gather)
1/4 teaspoon vitamin E oil
1/2 teaspoon pure bees honey
Mix all ingredients together and allow to sit in the sun for 1 full week. Strain off plant material and store perfume in the refrigerator, preferably in a glass container with a tight fitting cork or lid.

Henna flowers can also be used for medicinal and aromatic applications such as incense. Drying the henna flowers is said to make a wonderful way to scent your surroundings because the dried flowers hold their aroma for a good amount of time. Similar to rose petals. To dry the flowers, cut the branch below the flowers and below 2 sets of leaves. Strip of the leaves and tie the branches of flowers together. Such as in groups of 3. Then take a brown paper bag and cover the flower head. Allow the branches to stick out of the bag and tightly secure the paper bag around the exposed branches. Then take some twine and attach the branches to a hanger. Otherwise the flowers should be suspended from the hanger, so that the once open end of the bag is pointed upwards and the flower heads inside point down. This then needs to be placed in a very dry, dark and warm area of the house. The flowers can also be dried by spreading them on a pan and placing them into a very low temperature oven. Henna berries on the other hand are traditionally gathered in the fall in China. You can do the same when you see the little black berries appear. In some areas of the Middle East the berries were ground and used for dye. In India the seeds were made into oil via distillation, like grapeseed oil. On older trees, the bark can be peeled off, although much of the active properties of henna are really found in its leaves.

CUTTING THE MIGNONETTE TREE

> *"On the side of al-Masfal, which is at the far part of the town, is a mosque attributed to Abu Bakr the Faithful and has a lovely garden having palm trees, pomegranate and jujube trees and we saw hinna'trees."*
>
> —Ibn Jubayr, Mecca 1183

The best way to have more henna plants is to propagate in the *fall*. The reason is that for most, the fall is a time that is dryer. It is also a time when you should prune your henna bush and loosen the soil around the base anyway. Growth is very little in this time as well. Making it perfect for cuttings. The best place to grow and prepare your cuttings is inside a greenhouse or indoors using UV grow lights. When you are ready to make a cutting, look at the henna plant carefully. You should only cut branches that have produced blooms of henna flowers. These blooms should be spent (dried out). Count 4 sets of leaves down the branch (or stem) and cut it off cleanly with a sterilized knife. Remove the dried henna flowers. Then remove 2 sets of leaves from the bottom of the cutting, leaving 2 sets at the top. 1 inch from the bottom, using your sterilized knife, scrape of the outer bark very gently. You want the new green under part to show through. Then place the cuttings in water to soak. Fill a container such as a small terra-cotta pot or Styrofoam cup, making sure there are plenty of holes at the bottom for drainage. Fill the cups with loam or peat moss. Something very water retentive is needed. Next you will need either a self closing plastic bag or an empty large soda bottle. If using the soda bottle, take a sharp knife and cut off 4 inches from the top. Add a great deal of water to the soil in the cups and allow them to drain. Then make a hole, dip your cutting in a growth hormone product and place it into the hole you made. Take care to push the soil down around the cutting. Next water it again and allow it to drain. Take either the plastic bag or the soda bottle and place the cutting inside. Cover the soda bottle with plastic wrap or close the bag so that the moisture stays inside. Place the cuttings under grow lights or in a very sunny area of the green house. The temperature should stay around 70F degrees. Leave your cuttings be for 4 weeks. After this time, water them and look for new growth. Allow them to stay in the bag or bottle until the new growth becomes too much and they are cramped. In this time do not add any type of fertilizer product. They then can be transplanted into sterilized pots and grown as a tub plant or however you wish to use them. This method is called hardwood cuttings. Spring is also a time that allows for cuttings, which are called softwood cuttings. They are created in the same manner as hardwood cuttings. Typically healthy henna allows for 3 or 4 cuttings per year. Doing more may seriously weaken your plant and cause it to die.

In Victorian times, a topiary Mignonette Tree was quite in vogue. When the henna bloomed, it would create a sphere of wonderful aroma. Perfect for drowning out the awful smell of the dirty city street or country farm. You too can create Mignonette Trees with your henna (or cutting). This is very good for tub plants. In Europe the henna would be kept in large pots so that when it become too cold outside, the henna could be easily brought inside. In the fall, on a nice sized henna bush (3 feet or higher), find the center trunk of the plant. Hopefully, using supports if need be, your trunk is nice and straight. Using a sterilized knife or clippers, remove the branches starting from where the trunk meets the dirt up to about 12 inches from the top. What you are trying to create is a bare trunk and round sphere of foliage at the top. Then trim the remaining shoots so that they create a round shape. At first, due to a lack of density, the sphere may not look very round. Buy not to worry, come spring new leaves will start to grow and fill in the bare areas. You can then trim as your henna grows so that the sphere shape remains. To keep the trunk small, you may wish to wrap gardening wire around it. This sort of an arrangement takes a lot of care because the henna grows so quickly but would make for something different to do with your fragrant plants. One may also be able to use the graft method used for rose trees to create a Mignonette tree faster and taller.

Maintaining your henna bush is quite easy. In late spring you should remove dead branches (wood), bark and leaves. A good pruning is also in order at this time to promote growth. A little fertilizer specifically for deciduous (or bush / hedge type plants) can also aid in good growth. During fall, you should prune your henna bush again and collect the berries if you wish to propagate using seeds instead of cuttings. Loosen the soil at the base of your henna plant and if you anticipate the temperate dropping below 50F, bring you plant indoors. Leaf drop (of healthy leaves) can be due to the temperature being too cool, not enough water or not enough sunlight.

HENNA AND WORLD ECONOMY

> "[Giovanni] Mariti in his voyage to Syria and Palestine [1761] saw this shrub [henna] and its flowers in the region where the Blessed Mary traveled. The leaves are, in his opinion, smaller and more elegant than those of the myrtle; flowers, the color of pink, laid out by bouquets in the shape of bunches."
>
> —Anne Catherine Emmerich, 1813

Henna remains a driving force in poorer countries such as India and the Sudan. It is estimated that over 9,000 tons of henna is produced every year, with over 1,011 tons being imported to Cyprus, the UAE and Russia alone. Many countries that heavily produce henna also deplete their supply and are thus forced to import more from other smaller growing communities. It is hard to project the amount of henna being produced as many families grow it on a small scale as a cottage industry. Today the major growers of henna include Egypt, India, Pakistan, Iran and the Sudan. In Egypt, in ancient times, Uppor Egypt appeared to be the main source of henna production and exportation. Later, Nubia (around the Nile) became a major grower of henna for Egypt as well as for exportation, which was seen well into the Victorian era. Many writers in the 17th to 19th century noted Egypt and Africa as being the primary source for henna in both the Mediterranean and the Near East such as Turkey and Persia. Prior to the 1950's, India and Pakistan were one, so it is not surprising Pakistan remains a force behind large scale henna production. To date Rajasthan (Pali, Nagor and Sojat) and Haryana (Faridabad and Gurgaon) are the top states for producing quality henna in India. This is because these dry areas are the most well suited to growing henna. Other parts of India are subject to 3 month long rainy seasons which ruins the henna and causes it to suffer root rot. Not only this but moisture must not be introduced to the drying henna leaves as it can dramatically effect their ability to dye as a powder. Rajasthan (and surrounding areas) in general appears to be the oldest areas in India for henna farming, with many of the parcels being hundreds of years old. It was estimated India produced 7,600 tons of henna in 1993 and the amount is expected to continue to grow. The henna production in Sudan was beginning to fall but many female owned cottage farming industries have turned the trend around. Henna has and likely always will be used for pleasurable purposes. Its bitter taste and lack of nutrients prevented it from ever being used as animal fodder. The same is true for its use as food. Some have mistakenly confused it with Alkanet which was used in food but henna never has been used for such purposes because it is toxic if ingested. Therefore the main buyers of henna were women, for use as a cosmetic and as a medicine. If times were extremely hard, henna would not have been a priority like food or plants used for clothing. This is how we can see the very wealthy were the main ones buying and using henna. Similar to how chocolate was used by Aztec kings but today is enjoyed by many because of better farming and labor techniques. Henna needs to be processed, just like tea, in order to produce cosmetic and dye attributes. Without proper processing, it is impossible to transport henna to other locals. This processing is what drove the price of henna very high in ancient times. If a person lived in a city or was part of a nomadic tribe, they would be forced to buy henna. Likely at a special time in their life such as at their wedding or at a relatives death. It is hard for many to comprehend the worth of henna in ancient times because the main users today are the very poor in 3rd world conditions. This is a result of peoples abandoning traditional practices and trying to become more modernized, especially in the areas of medicine. Henna's use as a coolant, deodorant, hair colorant and skin healing preparation is what keep henna body art alive and henna on the market, even when fashions changed to more Western looks among the rich. Islamic people and Arabian traders in general were the main force behind henna use spreading, even as far as China. This is seen linguistically and how many names are based on the Arabic "hinna". The Fatimid period was a major time of henna growth and spread. A major crop well into the 12th century among Islamic peoples in the Near East and Mediterranean, it's easy to see how many non-Muslim women began using henna cosmetically. This included Christian Sicilian women being seen wearing henna designs to church. The major uses of the henna being produced however were both the bath houses and the women and men frequenting them. The culture of the bath house among Islamic people was a key source of income for henna growers and producers. During the Middle Ages, henna was produced in large quantities in Spain which was housed in huge vertical henna mills called "arha'al-hinna". Being a henna mill worker was seen as a very lowly job, likely because of the very bad respiratory problems that resulted (asthma, bronchitis, etc.). Many of these poor henna workers adopted Sufism, hence why henna is frequently found mentioned in Sufi poetry and so forth. Being an owner however was certainly a lucrative position, as many women in Spain and surrounding areas used henna to add color to their hair. Which they saw as a bleaching method. Seeing that Islamic people (especially Moors) were very attached to their henna and it was a bases for their economic power in Spain, the Spanish Inquisition made it a point to ban its use. This not surprisingly caused all of the henna producing mills in Spain to close and fall into disrepair. Spain never again became a major grower or producer of henna. The Qajar period of Persia documents wide scale henna production and mills. Due to Persia's great knack for trade, henna from the area could travel all about the Old World, including to China. While India appeared to have been producing large quantities of henna from at least the 16th century onward, it seemed that the country was consuming it itself and wasn't exporting it. Much of the henna, well into the Victorian era was still coming from Africa. India became a producer of henna, especially for England, with the 20th century movement to use it as a hair colorant. Australia and the UK are still today some of the top importers of henna powder. Henna is usually grown by farmers who also produce other goods. This is because, like orchards, henna must mature a year or two before a real harvest can be taken. The leaves are usually dried on the ground and then striped from the branches and taken to a market setting to be sold to mills. If the henna is of a very great quantity, brokers may be used. Singapore and Dubai remain major ports of henna trade. The whole henna leaves and stems are then milled, usually by large stone wheels. Sometimes oils are sprayed on the henna to keep the dust down. The henna then may be bleached or treated with other chemicals to make it stain better. A particularly dangerous one being PPD. Others include adding iron derivatives and solvents. Many mills, in an attempt to stay on top of competition, keep what they are doing to the henna a secret. It appears however that even as far back as the 1930's, henna has been found to be adulterated and contaminated with other substances. Which many times caused adverse health effects. Some mills even add green coloring to the henna powder in an attempt to make it look fresher. Henna is usually stored on the floor in huge heaps or in jute bags before being placed in recycled oil drums and exported on ships. This open air environment allows for contaminants to enter the henna. Many henna samples have been found tainted with salmonella &/or tetanus. Lead is also a major contaminant and has caused poisoning and neurological problems in people, especially in the Middle East. Many henna body art companies fail to do regular in house testing of the henna batches they obtain through imports. The US FDA has not approved henna for use on the skin, so they in affect do nothing to regulate it, except hold henna products with improper labeling &/or illegal color additives. Thus, at this time, one can not be certain if what they are getting is pure or not. One may find a measure of protection by buying whole, dried henna leaves and grinding them to powder as needed. Whole henna leaves were sold in the early 20th century for hair color but today, has been completely replaced by powdered henna products. Since henna is a botanical, it is not standardized. Otherwise, one batch may be dramatically different in its Lawsone (hennotannic acid) content which can make results vary. A study done by a dye crop researcher found that most henna mills do not extrapolate the Lawsone for measurement in order to ascertain the quality of the harvest. Even though many have the capability of doing so. Henna is relatively cheap by the ton. According to 1992 records, extremely high quality henna from Pakistan and India fetched $700 US dollars a pound. Low quality henna was sold for $250 a pound. Henna from Sudan fetched $500 a pound and Iranian henna got $300 a pound. One can easily see how companies reselling but 50 or 100 grams of henna a piece in the US for upwards of $5.00 a box would quickly make the money back.

Afterword

❀

The foregoing book was the result of much research, personal knowledge and meticulous preparation. It is the desire of the author to provide a book of the highest quality. Therefor feedback is encouraged, in preparation for future editions.

Please direct all materials and comments to the author of this book. All correspondence that includes a SASE or international reply coupon will receive a written response. The postal address is:

Marie Anakee Miczak
P.O. Box 312
Manalapan, NJ 07726 USA

www.anakee.com
mehndi@mehndi.zzn.com

About the Author

❀

Marie Anakee Miczak is the author of a number of titles including "Secret Potions, Elixirs & Concoctions" (Lotus Press) and "Mehndi: Rediscovering Henna Body Art" (BBOTW), both of which contain information on henna. Ms. Miczak has been trained in Aromatherapy with an AIYS (dip. aroma) from England based Kevala, herbology and the culinary arts in addition to a certification in Written English. Marie Anakee is Contributing Editor for the Aromatherapy topic with Suite101 and also teaches courses at Brookdale College, NJ. She has appeared in numerous publications including the New York Times and Writer's Digest magazine. Her official websites are http://www.miczak.com , http://www.mehndi.tajmahal.net and http://www.anakee.com. Ms. Miczak hopes to use henna/Mehndi for humanitarian purposes and has already raised money for UNICEF using her website Mehndi Mecca.

Appendix

Resources Concerning Henna:

Earth Harmony
P.O. Box 322
Norway, IA 52318 USA
1-888-301-7171 / 1-800-341-2604
FAX: 678-445-4996
www.frontier.com
Earth Harmony is a well known maker of essential oils sold in healthfood and newage stores. They also sell a wonderful *henna* attar which is made in India. It is simply wonderful! Earth Harmony is a sister company to the organic herbal company Frontier Herbs which sells henna powder.

Companion Plants
7247 North Coolvile Ridge Road
Athens, OH 45701 USA
phone: 740-592-4643
FAX: 740-593-3092
e-mail: complants@frognet.net
www.companionplants.com
One of the few greenhouses to grow henna plants which are ready to be mailed to your door (in the US). The plants (from cuttings) are great, grow fast and are under $10 each. Visit their website for more information.

The Hermit's Grove
P.O. Box 0691
Kirkland, WA 98083-0691 USA
www.thehermitsgrove.org
Founded by herbalist and author Paul Beyerl (The Master Book of Herbalism), offers 150 dried botanicals (including henna), monographs, educational programs, gardens and other resources for the medical and magical herbalist. For more information, send a SASE to the address above. His monographs are wonderful and the only ones on henna I could find.

Mehndi Mecca
www.mehndi.tajmahal.net
e-mail: mehndi@mehndi.zzn.com
The official website of this book and author Marie Anakee Miczak. Keep updated about henna and the research Ms. Miczak is doing on the subject. This website may also be accessed through www.miczak.com and www.anakee.com .

Other Miczak Titles:

Secret Potions, Elixirs & Concoctions
by Marie Anakee Miczak
Lotus Press 1999
Contains a section on henna and Mehndi body art as well as many natural recipes for health and beauty.

Natures Weeds, Native Medicine
by Dr. M. Miczak
Lotus Press 1999
A wonderful book filled with recipes and authentic Native American Indian wisdome. Herbs, foods, lore, illustrations, gardening and more.

Mehndi
by Marie Anakee Miczak
Infinity 1999
A fully illustrated book devoted to the art of henna—Mehndi. Recipes, step-by-step directions and more allows you to create henna body art in the comfort of your own home!

How Not to Kill Yourself with Deadly Interactions
by Dr. M. Miczak
Xlibris 2000
A must read book on the dangers of interactions between medications, foods, herbs, vitamins and more. Plus alternatives to more toxic prescriptions and lifestyle advice.

How Flowers Heal
by Marie Anakee Miczak
Writers Club Press 2000
Do you enjoy flower gardening? How about roses? Well did you ever think yo make a jelly, tea or soup from them, rich in vitamin C? Well this charming book allows you to learn all the ways flowers have been used throughout history for health, beauty and healing.

The Secret to Staying Young
by Dr. M. Miczak
Lotus Press 2001
Learn how both men and women of any age can use nature to help keep them healthy anf feel—looking young.

You can find any of these books at your local bookstore or you can buy them online at www.amazon.com , bn.com , borders.com , etc. For more information visit www.miczak.com or www.anakee.com. You can also e-mail miczak@juno.com or mehndi@mehndi.zzn.com .

Notes

❈

1.) According to copyright laws, all works published before the year 1923 are in the "public domain", which means works (or portions) may be used without the permission of the author. Likewise, the "fair use rule" allows for writers engaged in recounting history (in book form) to freely use quotes from various sources in order to comment on the material at hand. Facts are also not protected by copyright, especially in the area of history; provided that the information recounted is in ones *own* words, except if the work was published prior to 1923.

2.) For the most part the recipes have been recreated by the author *but* it is suggested that laypersons abstain from doing the same. The recipes provided in this book are for their historical worth only. Henna is likewise for external use only. The author nor the publisher takes any liability for the readers actions regarding the use of this book.

3.) Here is more synonyms for henna and their corresponding languages / countries. Names with [?] beside them mean they are unconfirmed by me personally.

Kina gibi (Turkish)
Kina yakmak (Turkish)
Kinalar yakmak (Turkish)
Danh to (Vietnamese)

KËNA (Albanian)
Pschapeddagoranta (Teluga) [?]
Makaracaka (Tamil)
Azavanam (Tamil)
Ponvannakkurijnci (Tamil)
Pitai (Tamil—also a name for tumeric)
Aivannam (Tamil)
Kurantakam (Tamil—also a name of conehead species)
Kuravakam (Tamil)
Korantam (Tamil—also a name of conehead species)
Kurajnci (Tamil—also a name for date palm tree)
Cakacari (Tamil—also a nickname for a 'wife', like sweety or honey)
Pavalakkurijnci (Tamil—also a name for crape myrtle)
Ponninpuvalamarudondri (Tamil) [?]
Kurijnci (Tamil—also a type of poem, conehead species, date plam tree)
Simru (Bhote) [?]
Krapin-kupin-tue [?]
Chi-gaip-hoa [?]
Khao-trien [?]
Kao-youak [?]
Mong-tay [?]
Bhurara (Lambadi) [?]

4.) Here you will find a list of organizations / charities helping women and children to have a fighting chance. Henna and Mehndi body art can be used as a way of shedding light on the many problems women face, especially in 3rd world countries. The below have been collected for informational purposes,

I (Marie Anakee) nor the publisher are endorsing or validating them. Give carefully and always fully investigate prior to doing so. Visiting the org.'s website is a great start!

Forward
RT# 292403
6th Floor, 50 Eastbourne Terrace
LONDON
United Kingdom
W2 6LX
Telephone: +44 (0)207 725 2606
Fax: +44 (0)207 725 2796
www.forward.dirdon.co.uk
E-mail: forward@dircon.co.uk
Founded by Efua Dorkenoo to tech and eradicate the practice of FGM in the UK. Visit the website for a wealth of information.

UNICEF—United Nations Childrens Fund
www.unicef.org
e-mail: netmaster@unicef.org
Helping children in poor areas of the world and keeping an eye on Bride Burning, child abuse and adverse health conditions.

Rising Daughters Aware Website
www.fgm.org
e-mail: director@fgm.org
A very informative website for lay people and people in medicine. Many links to websites working to stop FGM can also be found here.

The FMG Education & Networking Project
P. O. Box 6597
Albany, CA 94706 USA
phone: 510-558-1012
FAX: 603-853-7789
www.fgmnetwork.org
e-mail: fmg@fgmnetwork.org

Very informative website with many links to further reading on the subject.

ISADABBI—International Society Aganist Dowy & Bride-Burning in India, Inc.
P.O. Box 8766
Salem, MA 01971 USA
e-mail: hthakur@shore.net
Providing education and prevention of the dowy and Bride Burning muders in India. UNICEF now notes that over 25,000 women have been burned alive this past year. The numbers are rising every year! Yet no one ever goes to jail for the crimes.

Glossary

❀

Akkadian—Language and writing system belonging to the Semitic subfaimly called Assyro-Babylonian. It was used from 3000 BCE until the time of Christ.

Arabic—A member of the Semitic subdivision of Hamito-Semitic faimly of languages and spread with the teachings of Islam.

Aramaic—A Hamito-Semitic language that was used in the Fertile Crescent and Syria around the time of Christ. Henna's name in Aramaic was derived from the Hebrew "ko'pher".

Astringent—A plant produced substance that coats a surface thus prtecting it. Also a contracting and binding effect.

Chakra—Believed in Ayurvedic medicine to be both the energy and spiritual power points of the body.

Emollient—Something that softens and soothes the skin.

Essential Oils—Violatile oils / substances extracted from plants. Many times they are highly aromatic and contain over 100 chemicals.

Floristic—Flowering Plants.

Floristic Kingdom—A faimly of plants and an area of the globe where they are thought to have originated.

Galenical—Traditional system of medicine formed in ancient Greece by Galen.

Hebrew—A Semitic of the Hamito-Semitic subdivision.

Humor—Theoretical body fluid that is a part of Ayurvedic and Galenical medicine. Henna is related to Black Bile and thus the Melancholic humor.

Lawsone—The main coloring chemical in henna and also one of its active ingredients. In the living leaf Lawsone aids in henna's photosynthesis of light into food.

Mehndi—Meaning myrtle in Hindi and Sanskrit and later henna as well.

Sedative—Something that is soothing and calming. At times helping with sleep. Henna, especially its essential oil is sedative in nature.

Tang Period—From 618 to 907. Likely when henna was taken by China to Japan.

Tannin—Astringent, it is an active ingredient of plants and combines with proteins like Lawsone. It aids henna in coloring the skin and tanning.

Taxonomy—Classification and study of origins of various plant species. Henna began to be studied in this way in the 18th century and likely first in India.

References

Chapter 1:
Albert Y. Leung, "Chinese Healing Foods and Herbs" (AYSL Corporation, NJ: 1984) 62-66.
E. Cobham Brewer, "The Dictionary of Phrase & Fable: New & Enlarged Ed.", (1894).
Hippocrates, "On Ulcers" (_____: 400 BCE).
Homer, "Iliad" (750 BCE).
James Duke, "Botanical Database".
John H. Young, "Our Deportment, or the Manners, Conduct, and Dress of the Most Refined Society; including Forms for Letters, Invitations, Etc., Etc. Also, Valuable Suggestions on Home Culture and Training. Compiled from the Latest Reliable Authorities" (F.B. Dickerson & Co., MI: 1883).
John Gill, Scripture Notes, 17th century
John Wesley, Scripture Notes, 18th century
Kate Greenaway, "Language of Flowers" (1885).
Lise Manniche, "Sacred Luxuries: Fragrance, Aromatherapy & Cosmetics in Ancient Egypt" (Cornell University Press—NY: 1999).
Ms. M. Grieve, "A Modern Herbal" (1931).
Mary Chauancy, "The Floral Gift: from Nature & the Heart" (Leavitt & Allen, NY: 1853).
M. G. Easton, "Easton Bible Dictionary" (Thomas Nealson: 1897)

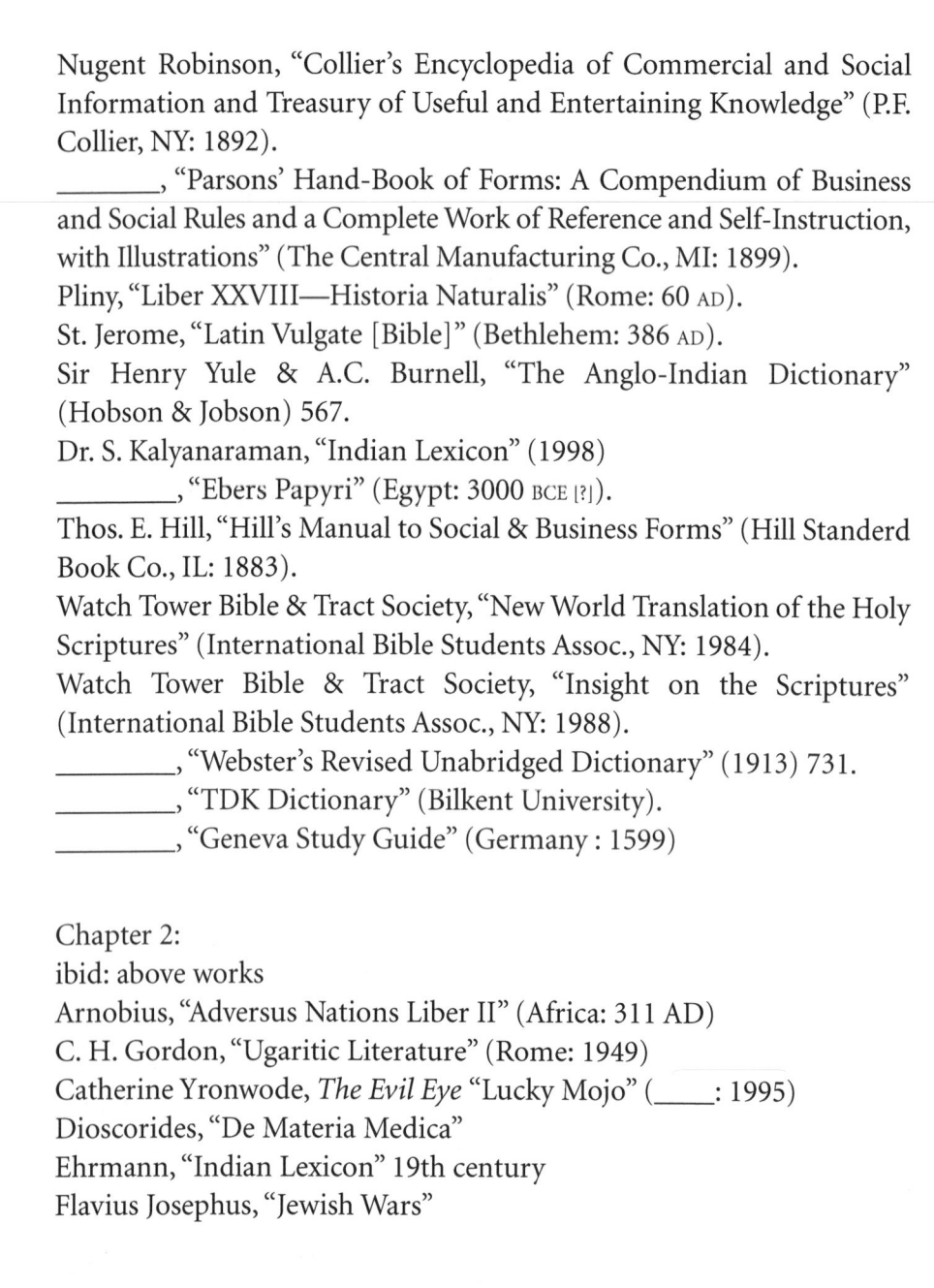

Nugent Robinson, "Collier's Encyclopedia of Commercial and Social Information and Treasury of Useful and Entertaining Knowledge" (P.F. Collier, NY: 1892).
_____, "Parsons' Hand-Book of Forms: A Compendium of Business and Social Rules and a Complete Work of Reference and Self-Instruction, with Illustrations" (The Central Manufacturing Co., MI: 1899).
Pliny, "Liber XXVIII—Historia Naturalis" (Rome: 60 AD).
St. Jerome, "Latin Vulgate [Bible]" (Bethlehem: 386 AD).
Sir Henry Yule & A.C. Burnell, "The Anglo-Indian Dictionary" (Hobson & Jobson) 567.
Dr. S. Kalyanaraman, "Indian Lexicon" (1998)
_____, "Ebers Papyri" (Egypt: 3000 BCE [?]).
Thos. E. Hill, "Hill's Manual to Social & Business Forms" (Hill Standerd Book Co., IL: 1883).
Watch Tower Bible & Tract Society, "New World Translation of the Holy Scriptures" (International Bible Students Assoc., NY: 1984).
Watch Tower Bible & Tract Society, "Insight on the Scriptures" (International Bible Students Assoc., NY: 1988).
_____, "Webster's Revised Unabridged Dictionary" (1913) 731.
_____, "TDK Dictionary" (Bilkent University).
_____, "Geneva Study Guide" (Germany : 1599)

Chapter 2:
ibid: above works
Arnobius, "Adversus Nations Liber II" (Africa: 311 AD)
C. H. Gordon, "Ugaritic Literature" (Rome: 1949)
Catherine Yronwode, *The Evil Eye* "Lucky Mojo" (____: 1995)
Dioscorides, "De Materia Medica"
Ehrmann, "Indian Lexicon" 19th century
Flavius Josephus, "Jewish Wars"

Galen, "medical treatises" (____:180 AD)
Henry Field, "Southern Asian Body Markings" (Peabody Museum: ____)
Henri Gamache, "Protection from Evil" (____: 1946)
Joanna Waley-Cohen, "The Sextants of Beijing: Global Currents in Chinese History" (WW Norton & Co.: ____)
John Fryer, "Fryer's India and Persia" (____: 1681)
Theophrastus, "On Odors" (____: 285 BCE)
_____, "Papyri Ebers" (Egypt: 1500 BCE)
_____, "Leyden and Stockholm Papyri" (Egypt: 250 AD)
_____, "Karma Sutra" (India: 320—540 AD)
_____, "Ugaritic Handbook" (Syria: 1935)
_____, "The Shrine of Saft-el-Henneh & the Land of Goshen" (____:____)
_____, "Brewer's Phrase & Fable"

Chapter 3:
Sir Samuel W. Baker, "Baker's Nile Tribes" 19th century
Ann Berwick, "Aromatherapy A Holistic Guide: Balancing Body and Soul with Essential Oils" (Llewellyn, MI: 1996)
A. Lucas, "Ancient Egyptian Material & Industries" (____ London: 1962)
D. M. Stoddart, "Perfumes" (____:____)
Herodotus, (____ : 425 BCE)
L. Foreman, "Cleopatra's Place: In Search of a Legend" (Radom House—NY—1999) 61.
John Lightfoot, "from the Talmud and Hebraica" (____: 1650)
Julia Lawless, "The Illustrated Encyclopedia of Essential Oils: The Complete Guide to the Use of Oils in Aromatherapy and Herbalism" (Element Books, MA: 1995)
Lise Manniche, "Sacred Luxuries: Fragrance, Aromatherapy & Cosmetics in Ancient Egypt" (Cornell University Press—NY: 1999).

Sus'ruta Samhita (India : 400 AD [?])
William Loring, "A Confederate Soldier in Egypt"
_____, "The Garden of Latin Health" *Translated in French*

Chapter 4:
Lord Byron, "Don Juan" (___: 1819)
Sir Richard Burton, "Pilgrimage to Mecca" (England : 1859)
Baron Dominique Vivant Danon, Ancient Egypt, (France: 18th century)
Dadaloglu, Poems, (Turkey : 1850)
Dallana, (India: 1100 AD)
Dominique Vivant Denon, "Ancient Egypt" 18th century
Dr. Lang, "Nuremberg Journal", (Germany: 1526)
H. Rider Haggard, "Smith & the Pharaohs" (___: 1912)
Imr-Ui-Quais, "Hanged Poems", (Ka'aba Mecca: 622 BCE)
Parakarana II, "Navanitaka" (India : 2nd century AD)
Samuel G. Wilsom, "A Year Amoung the Persians" (___: 1895)
_____, "A Year Amoung the Persians", 20th century

Chapter 5:
_____, "Mehndi: Rediscovering Henna Body Art" (BBOTW, PA: 1999)
ABCR, "2-Hydroxy-1, 4-naphthoquinone" (ABCR Gmb H & Co. KG).
Albert Y. Leung, "Chinese Healing Foods and Herbs" (AYSL Corporation, NJ: 1984) 62-66.
Anand, K. K. *et al*, "Planta Medica" (___:___) 22
Avicenna, "The Canon of Medicine" (Persia: 1030)
BODD (the Botanical Dermatology Database) "Lawsonia inermis".
ChemFinder.com, "ChemFinder Online Database [search word henna]".
Dr. Shaw, 18th century
Ebers Papyri, (3000 BCE [?]).

Environmental Science Center, "PhysProp Database".
Harvey Wickes Felter, MD & John Uri Lloyd, P.hD., "King's American Dispensatory: 18th Ed." (1898).
James Duke, "CRC Handbook of Phytochemical Constituents of GRAS herbs and other Economic Plants" (CRC Press, FL: 1992).
James Grey Jackson, (____: 1820)
John Fryer, "Fryer's India and Persia", (____: 1681)
Lesley Bremness, "Herbs: The Visual Guide to More Than 700 Herb Species from Around the World" (DK, NY: 1994) 111.
Lise Manniche, "Sacred Luxuries: Fragrance, Aromatherapy & Cosmetics in Ancient Egypt" (Cornell University Press—NY: 1999).
Ms. M. Grieve, "A Modern Herbal" (1931).
M. S. Zevada, *The Historical Use of Henna in the Balkins—Thaiszia* _____, (____: 1993)
Michael McGuffin, Christopher Hobbs, Roy Upton & Alicia Goldberg (Editors), "Botanical Safety Handbook" (CRC Press, NY: 1997).
Nicholas Culpeper, "The Complete Herbal" (1653).
Nist, "Chemistry WebBook [search word henna]".
NTP, "Chemical Health & Safty Data [Sheets]".
Paul V. Beyerl, "Henna: Lawsonia alba Lawsonia inermis Linn. Lythraceae Monographs" (The Hermit's Grove, WA, ____).
_____, "PDR for Herbal Medicine" (Medical Economics Company, NJ: 1998).
Penelope Ody, "The Complete Medicinal Herbal: A Practical Guide to the Healing Properties of Herbs, with more than 250 Remedies for Common Ailments" (DK, NY: 1993).
Proton NMR Spectral Molecular Formula Index [henna] (NMR).
Stefan Chmelik, "Chinese Herbal Secrets: The Key To Total Health" (Avery, NY: 1999) 22-37.
Int Pediatr. 2000;15(2):114-116

_____, "The Garden of Latin Health" *Translated in French*
_____, "Encyclopedia of Religion and Ethics" (____: 1910)

Chapter 6:
C. S. Sonnini, "Travels in Upper and Lower Egypt" (France : 18th century)
Havelock Ellis, "Studies of Sexual Psychology" (France: 1935)
Homer Smith, "Man and His Gods" (NY USA : 20th century)
Kalidasa, (India : ____)
Thalaba, Poetry, (____: 1800)
_____, "Karma Sutra" (India: 320—540 AD)
_____, "The Perfumed Garden" (____: 900 AD)

Chapter 7:
C. H. Ebermayer, "Handbook of the Pharmacist" (Germany: 1821)
Indian Standards Insttute, "Specifications of Henna Powder" (India: 1984)
Ruth Winter, "A Consumer's Dictionary of Cosmetic Ingredients: Complete Information about the Harmful & Desirable Ingredients Found in Men's & Women's Cosmetics" (____, ____) 118.
Segundo Gibaja, "Natural Pigments" (South America:____)
_____, "18th Century British Terms"

Chapter 8:
Alfonso X. Elsabio, "The Book of Chess, Dice and Board Games" (Spain: 1282)
Charles Sigisbert Sonnini, "Voyage to Syria" (___: 1717)
J.M. McGarvey, "Lands of the Bible" (____: 1880)
Lucy Aikin, Poetry, (___: 1810)

Muhammad Din Tilai, Poetry, 19th century
Theophilus, 12th century
W. Walther, "Women in Islamic History: From Medieval to Modern Times" (Markus Weiner Publications, Princeton NJ: 1993).
Yazid Bbn Moavia, Poetry, (Arabia : 600 AD)

Chapter 9:
Adam Clarke, Scripture Notes, 1810
Buck Whaley, "Voyage to Turkey" (Irland: 1797)
Buffy Johnson, "Lady of the Beasts" (Harper Collins: 1988)
Jamieson, Faussett & Brown, Scripture Notes, (____: 1871)
Michaelis, Scripture Notes, (____: 1709)
Shabeni, "Description of Timbuktu" (____: 1787)
William Loring, "A Confederate Soldier in Egypt" (____: 1877)
_____, "Mehndi: Rediscovering Henna Body Art" (Infinity, PA: 1999)
_____, "Banner of the Arahants" (____:____)
_____, "Muslim, Hindu, Sikh Violence" (____: 1940)
_____, "Dede Korkut"

Chapter 10:
Forbes, (____: 1813)
_____, "Soviet Antropology and Archeology—Dede Korkut" (USSR:____)
_____, *The Perfumed Mummy*, _____ (____:____)

Chapter 11:
Carles F. Horne, "The Sacred Books & Early Literature of the East" (____, NY: 1917)
Gustave Flaubert, "Stammbo" (France: 1862)

Pat Matusky, *Types of Music in Malaysia—Gendang Tari Inai*, _____ (____: 1997)
Sri Guru Granth Sahib, "Sikh Holy Book" (India:____)
William Lane, "An Account of the Manners and Customs of the Modrn Egyptians" (London England: 1836)

Chapter 12:
Ann Killigrew, Poetry, (___ : 1685)
Imr-Ul-Kais, "Moallakat" (___: 540 AD)
Mr. John French, "Art of Distilation" (England: 1651)
Soranus, 1st century AD
St. Jerome, "Latin Vulgate Bible" (Turkey: 300 AD)
Sir Thomas Browne, Scripture Notes, 17th century
Cato Elder, "On Farming" (234 BCE)
Francois Rabalais, "Le tiers livre des faicts et dicts Heroiques du bon Pantagruel" (France: 1552)
John Batist Porta, "Natural Magick", 17th century
John Gerard, "The Herball" (England : 1597)
Jonathan Warren, *Reflections on an Ellectric Scribe*, _____ (___:___)
Nicolas Culpeper, 17th century

Chapter 13:
Anne Catherine Emmerich, "Life of the Blessed Virgin According to Anne Catherine Emmerich" (Germany: 1854)
Homer Smith, "Man and His Gods" (___:1950)
Saksena, Jogendra, "Art of Rajasthan" (India: 1979)
William Lane, "An Account of the Manners amd Customs of the Modern Egyptians" (London England: 1836)

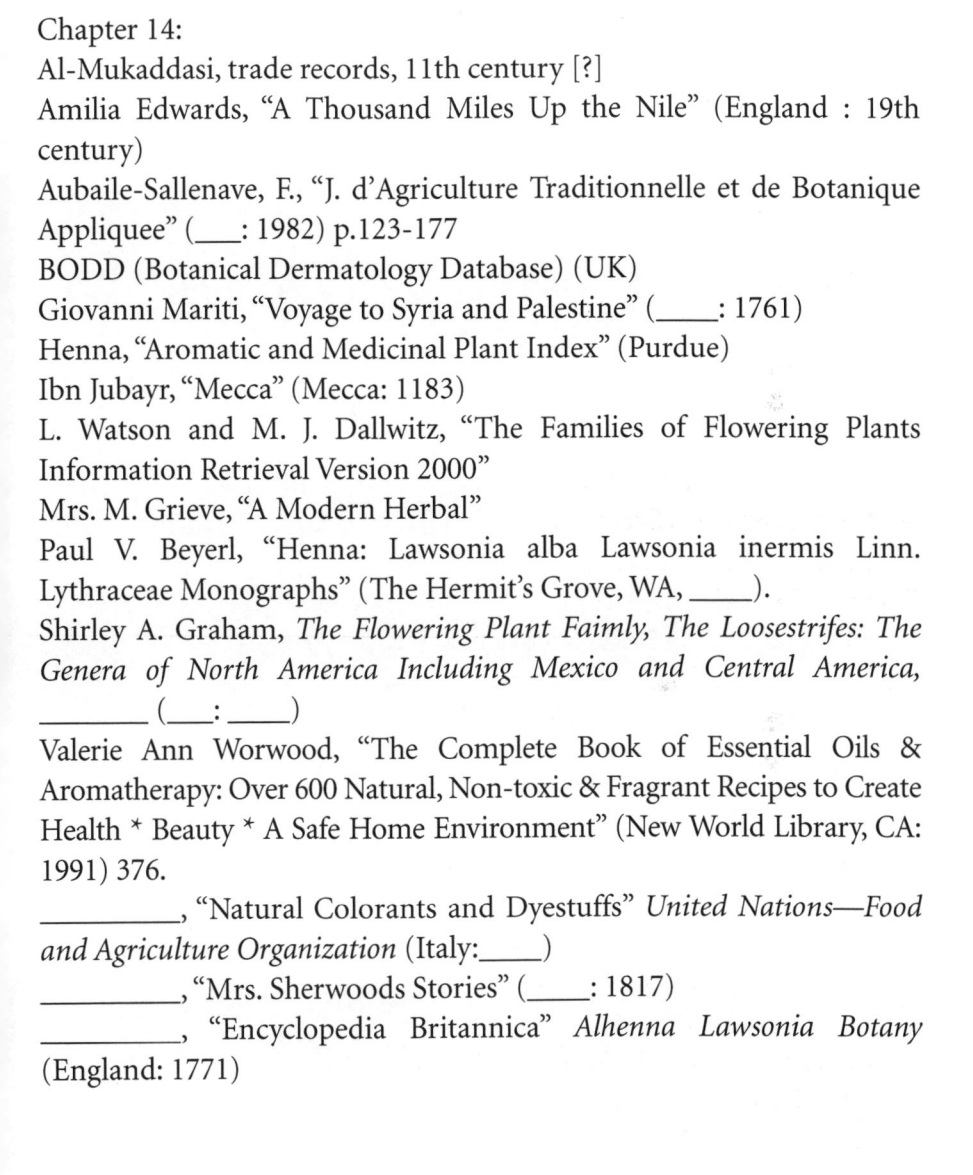

Chapter 14:
Al-Mukaddasi, trade records, 11th century [?]
Amilia Edwards, "A Thousand Miles Up the Nile" (England : 19th century)
Aubaile-Sallenave, F., "J. d'Agriculture Traditionnelle et de Botanique Appliquee" (___: 1982) p.123-177
BODD (Botanical Dermatology Database) (UK)
Giovanni Mariti, "Voyage to Syria and Palestine" (___: 1761)
Henna, "Aromatic and Medicinal Plant Index" (Purdue)
Ibn Jubayr, "Mecca" (Mecca: 1183)
L. Watson and M. J. Dallwitz, "The Families of Flowering Plants Information Retrieval Version 2000"
Mrs. M. Grieve, "A Modern Herbal"
Paul V. Beyerl, "Henna: Lawsonia alba Lawsonia inermis Linn. Lythraceae Monographs" (The Hermit's Grove, WA, ___).
Shirley A. Graham, *The Flowering Plant Faimly, The Loosestrifes: The Genera of North America Including Mexico and Central America*, _____ (___: ___)
Valerie Ann Worwood, "The Complete Book of Essential Oils & Aromatherapy: Over 600 Natural, Non-toxic & Fragrant Recipes to Create Health * Beauty * A Safe Home Environment" (New World Library, CA: 1991) 376.
_____, "Natural Colorants and Dyestuffs" *United Nations—Food and Agriculture Organization* (Italy:___)
_____, "Mrs. Sherwoods Stories" (___: 1817)
_____, "Encyclopedia Britannica" *Alhenna Lawsonia Botany* (England: 1771)

Bibliography

Mehndi: Rediscovering Henna Body Art
by Marie Anakee Miczak
ISBN 0-7414-0280-7, 200 pages (Infinty 1999) (1-800-BUYBOOK)
This book (by me) is for anyone that would like to focus more on the artistic side of henna. Fully illustrated, it walks you through creating all sorts of body art designs. Full recipes and expert notes make this a must read. For more information you can visit www.mehndi.tajmahal.net or www.anakee.com .

A Trade Like Any Other: Female Singers and Dancers in Egypt
by Karin Van Nieuwkerk
ISBN 0-2927-8723-5, 226 pages (Univ. of Texes 1995)
A must read book on belly dance as a whole.

An Ancient Egyptian Herbal
by Lise Manniche
ISBN 0-2927-0415-1, 176 pages (Univ. of Texes 1989)
Really the only book covering the medicinal herbs used by the ancient Egyptians.

The Henna Page
www.hennapage.com
e-mail: jr@hennapage.com

Started in 1997 by Jeremy Rowntree to be a guide and forum on henna body art, it has many areas of interest including a gallery of designs and PPD warning page.

An Illustrated Encyclopadia of Traditional Symbols
by J. C. Cooper
ISBN 0-5002-7125-9, (Thames & Hudson 1987)
Many do not understand that there is no such thing as a "Mehndi" or henna looking design. Many henna designs are traditional symbols that hold the same meanings, as they would for weaving, floor design and so forth. Thus this book is very helpful in bringing forth the meaning of henna's designs all over the Old World.

Sacred Luxuries
by Lise Manniche
ISBN 0-8014-3720-2 (Cornell Univ. Press 1999)
Wonderful book with a lot of space devoted to henna perfumes! (of course nothing that can not be found in this book but if you would like to see nice photographs, seek this book out.)

Index

❀

A

Africa, xxii, 9-11, 26, 37, 49, 59, 62-63, 65, 93-94, 99, 117, 132, 143, 145-146, 149, 152, 180, 182, 195, 199, 201, 203, 211, 214, 232, 262, 266, 269, 275-276, 290, 293, 310
Ajanta, xxiv, 53-55, 195, 203, 206
Akkadian, 4, 18, 25, 28, 34-35, 64, 201, 307
Alba, xxi, 6-9, 12-13, 84, 126, 133, 168, 181, 186-187, 194-195, 239, 273-274, 278, 313, 317
Alkanet, 12, 15-17, 19, 30-32, 94, 97, 100, 103, 106, 108, 135, 144, 158, 163, 198, 203, 208-210, 240, 251-255, 259, 291
Anatolia*****
Aphrodisiac, xxi, 14, 45, 49, 82, 85, 122, 124, 144, 155, 158, 160, 162, 215, 220
Arabic, 2-6, 8, 12, 16, 19, 35, 49, 63-64, 78, 100, 185, 193, 201, 251, 292, 307
Arabians, 99, 127
Archeology, 225, 315
Aromatherapy, 80, 82-85, 144, 297, 309, 311, 313, 317
Art, xi, xxiv-xxviii, 2, 7, 24, 44, 51, 53-55, 58, 62-64, 74, 86, 96-97, 138, 146-147, 150, 152, 164, 166, 168-172, 174, 176, 180-184, 186-188, 192-198, 200-211, 213, 215-218, 224, 227, 233-234, 238, 240-241, 247, 264-266, 278, 292, 294, 297, 300-301, 304, 312, 315-316, 319-320
Artifacts, 25-26, 28, 34-35, 41-42, 56, 59, 63-64, 87, 89, 110, 115, 238, 240-242, 276
Ayuveda*****

B

Beauty, cosmetic uses, xxiv, 103
Berries, 8, 12, 43-44, 72, 88, 97, 105, 129, 133, 135-137, 177-178, 209, 256-257, 259, 275, 287, 290

Bible, xxv, 11, 19-21, 25, 28, 32, 38, 45-46, 74, 76, 107, 161, 183, 212, 214, 243-245, 247-248, 250, 309-310, 314, 316

Body art, xxiv-xxviii, 7, 24, 44, 51, 54, 58, 62-63, 74, 86, 138, 146-147, 150, 152, 164, 166, 168-172, 174, 176, 180-184, 186-188, 192-198, 200-206, 208-211, 213, 215-218, 224, 227, 238, 240-241, 265-266, 278, 292, 294, 297, 300-301, 304, 312, 315, 319-320

Botany, xxii, 14, 23-24, 59, 273, 317

Buddhist, 54, 67, 137-138, 198, 203, 206-207, 213

C

Camphire, 19-20, 215, 243-250, 252, 255

Camphor, 20, 82, 144, 159, 243, 245-250, 259

Canaanite, xvii, 27-32, 37

China, 11, 30, 50-53, 60, 65, 98, 100-101, 105, 118, 127, 133-138, 154, 206, 210, 247, 249-250, 266, 276, 278, 280, 286-287, 292-293, 308

Christian, 127, 160, 212-215, 225, 250, 292

Cultivation, 272

Coffer*****

D

Death, xv, 7, 30, 78, 87, 90, 103, 109, 120, 146, 230, 232, 256-257, 261-262, 270, 291

Delta, Egypt, 37, 61, 119, 171, 276-277

E

Egypt, xxv, 3-4, 6, 11, 14, 21, 25, 27, 30, 37-38, 40-44, 46, 48, 53, 59, 61, 68, 70, 73, 76-78, 84, 88-90, 93-94, 97-99, 104-105, 107, 109, 114-115, 117, 119-120, 127, 135, 160-161, 168, 183, 185, 202, 209-210, 212, 215, 220, 222, 224, 230-231, 237, 256, 275-277, 279, 290, 309-315, 319

Egyptology*****

Essential oil, xxv-2, 24, 49, 67, 70, 76, 78-85, 93, 114, 116, 129, 151, 156, 158-159, 161, 176, 178, 285, 308

F

Farsi, 3, 5, 10

Fertility, 34, 80, 109, 132-133, 145, 153, 159, 163-165, 261, 265

Floristic, 23, 276, 307

Index 323

Flowers, henna, xxv, 10, 16, 44, 67, 70-72, 76, 79, 81, 83-85, 98, 104, 106-107, 109, 135-137, 141, 157, 159-162, 179, 206, 215, 232, 239, 248, 271, 279, 286-288

G

Garden, 52, 109, 111, 135, 159, 165, 198, 223, 259, 272, 279-281, 287, 312, 314
Greek, 3-5, 8, 13-14, 16, 19, 32, 35, 44, 46, 49, 67-70, 72, 74, 100, 119, 123-124, 128, 138, 200, 208-209, 240, 248, 251, 254
Gulhinna, perfume, 79, 84

H

Hair, 12, 26, 39-41, 48, 50, 52, 57, 59-60, 67, 75, 85-98, 100, 102-103, 108-109, 117, 119, 122, 127-128, 135, 138, 141, 143-145, 147, 149, 156-157, 167-170, 180, 182, 188-189, 193-194, 196, 199, 201, 204, 211, 215-217, 221, 226, 228, 230-231, 238, 241-242, 254, 267, 292-294
Harvesting, 18, 136, 256, 272, 285
Hebrew, 3-4, 8, 13, 17-18, 21, 28, 35, 45-46, 64, 74, 244, 307-308
Hindu, 21, 55-56, 152, 198-199, 203, 205, 213, 225, 315
Hinna, 2-3, 5-6, 12, 16, 35, 38, 84, 193-194, 201, 220, 278, 287, 292

I

India, xxiii-xxv, xxviii, 7-8, 11, 13, 16-17, 21, 24, 37, 43-44, 51, 53-58, 63, 66, 72, 78-80, 83, 89, 91, 93, 101, 110, 112, 114, 117, 120, 127-129, 132-135, 137, 142, 147, 149-150, 152-154, 156-157, 159, 162-163, 168, 170-171, 180-181, 183, 189, 193-199, 201, 203-208, 210, 218, 221-225, 227-228, 230, 232, 235-236, 239, 251-252, 255-256, 259, 262-265, 276, 278, 281, 284, 286-287, 290-291, 293-294, 299, 306, 308, 311-314, 316
Islam, 54, 63, 65, 90, 100, 121, 199, 201, 203, 216, 307
Israel, 17, 76, 83, 91, 224, 251, 254

J

Japan, 9, 51, 138, 249, 278, 308
Jews, 4, 17, 21, 29, 36, 45-46, 57, 76, 90, 102, 108-109, 118, 213, 221, 223-225, 257

K

Karma Sutra, 52, 311, 314
Kingdom, floristic, 23, 276, 307
Ko'pher, 3-4, 8, 13, 19, 21, 28, 33, 45, 74, 214, 244, 248, 307
Koran*****

L

Latin, xxi, 2, 7-8, 11, 16, 19-21, 46, 50, 68, 127, 158, 214, 243-246, 248-249, 251, 259, 274, 310, 312, 314, 316

Lawsone, 7, 52, 101, 104-105, 135, 146, 149-150, 153, 169, 173-175, 177-178, 216-217, 242, 284, 294, 308

Lawsonia alba, 6, 12-13, 168, 181, 186-187, 194, 313, 317

Lawsonia inermis, xxi, 2-3, 7, 12, 14, 16, 40-41, 52, 69, 71, 84, 157, 160, 167-168, 182, 185, 194, 239, 247, 273-274, 278, 312-313, 317

Lawsonia ruba, xxi, 7, 84, 98, 107, 239, 274

Lawsonia spinosa, 270

Linguistics, 35, 195, 241

Lythraceae, 16, 273-274, 278, 313, 317

M

Medicine, xvii-xviii, xxi-xxii, xxv, 14-15, 26, 41, 46-47, 49-50, 68-69, 72, 83, 85, 104, 114-125, 127-130, 132-139, 141, 143-146, 175, 200, 259-260, 291-292, 300, 305, 307-308, 312-313

Mediterranean, xxiii, 26-29, 31, 35, 43, 61, 74, 77, 93, 156-157, 201, 214, 251, 254, 277, 290, 292

Mehndi, xviii-xix, xxi, xxiii-xxv, xxvii-xxviii, 8-9, 17, 22-24, 55-56, 79, 84, 86, 96, 98, 101, 105, 113, 152-153, 168-169, 171-172, 180-182, 189, 191-194, 197, 199, 204-206, 208-210, 216, 218, 227, 251, 255, 262, 264, 295, 297, 300-301, 304, 308, 312, 315, 319-320

Mesopotamia, 4, 24-25, 36, 115

Mill, 148, 292

Mimata, 7, 84, 239, 274

Mummy, 27, 38, 40, 87, 98, 109, 231, 238-239, 315

Myrtle, xxiii, 6, 17, 19, 47, 53, 56, 74-75, 158, 189, 193-194, 198, 208, 255-260, 274, 290, 304, 308

Myth, 106, 115, 163, 203-204, 209-210, 259-260, 286

N

Night of the henna, xxiii, 57, 220-225, 235-237

Nile, 26, 37-38, 59-60, 68, 70, 75, 87, 90, 156, 165, 268, 276-277, 279, 281-282, 290, 311, 317

Notes, 6, 8, 25, 47, 55-56, 60, 70, 74, 92, 101, 120, 144, 186, 194-195, 214, 228, 246, 248, 270, 303, 306, 309, 315-316, 319

Nubia, 42, 59, 90, 276-277, 279, 290

O

Oil, xviii, xxv, 1-3, 11, 14, 17, 19, 24, 26, 28, 32-33, 42-44, 46-47, 49, 60, 67, 69-73, 75-85, 89, 91, 93, 98, 102, 104, 106-110, 114, 116-117, 119, 122, 129, 144, 148, 151, 156, 158-161, 168, 176, 178, 181, 193, 217, 221, 224-225, 230, 240, 242, 248-250, 255, 257, 259, 285-287, 293, 308

Ointment, 3, 19, 33, 43-44, 60, 62, 72, 76, 91, 96, 119, 158, 252-253

P

Pagan, 31, 116, 199

Papyrus, 40, 68, 115-116

Perfume, xxi, xxv, 11, 16, 24, 28, 34, 42, 44, 48-49, 57, 67-68, 70-81, 83-84, 98, 106, 109, 114, 156, 158, 161, 178, 193-194, 200, 215, 227, 230, 259, 270-271, 278, 286

Persia, 3, 5, 24, 51, 59, 92, 101, 108, 120, 123, 137, 180-181, 183-184, 186, 201, 211, 290, 293, 311-313

Phytomedicine*****

Pharmacology, xiii, xvii

Poem, xxi, 36, 50, 90, 157, 182, 184, 250, 304

Q

Queen, Cleopatra, 11, 47, 63, 67-68, 70-71, 75, 78, 80, 90, 97, 105-106, 156-158, 311

R

Red, ocher, 107-108, 268

Red sea, 53, 68, 83

Ritual, xxiii, 3, 33, 70, 142, 230, 269

Roman, 18, 40, 44, 68, 78, 90, 92, 98, 120, 123, 156, 230, 240

Ruba, xix, 7, 84, 126, 274, 278

S

Samerian*****

Seeds, 4, 8, 34, 43-44, 53, 62, 71-73, 79, 88, 129, 133, 135-137, 156, 177-178, 254, 273, 275, 277, 281-282, 287, 290

Solar Sringar, 197

Solomon, King, ix, xv, 19, 21, 34, 45, 76, 156, 245, 279

Spain, 11, 59, 99, 103, 185-186, 224-225, 270, 292, 314
Spinosa, xxi 7, 84, 98, 107, 239, 270, 274, 278
Sudan, 171, 269, 276-277, 290-291, 294

T
Tablet, 36-37
Tibet, 207
Tinctoria, 7
Turkey, 5, 11, 31, 35, 92, 141-142, 165, 201, 221, 231, 239, 244, 251, 254, 290, 312, 315-316

U
Unini tib—tibb*****

V
Vermilion, 12, 25, 229, 235
Victorian, xxiii, 4, 9, 14, 40, 59, 98, 107, 259, 267, 272-273, 277, 279, 289-290, 293

W
Wedding, xxiii, 63, 65, 102, 204, 220-228, 230, 237, 257, 262-263, 291

Y
Yemen, 11, 84, 200

Z
Zone, 280

Printed in Great Britain
by Amazon.co.uk, Ltd.,
Marston Gate.